Anna Elisabeth Höhl
Scientific Understanding



**Philosophy – Enlightenment – Critique** | Volume 2

## Editorial

The book series **Philosophy – Enlightenment – Critique** aims to promote philosophical thinking dedicated to a future worth living for everyone in times of global crisis. Climate change, political and religious authoritarianism, and growing social inequalities – the manifold and interconnected issues of our time require a return to the power of reason. In the spirit of a new enlightenment, the series initiates a dialog between different philosophical schools and traditions that critically scrutinize the past and present and explore the ramifications of sustainable alternatives. This requires a re-examination, re-interpretation and revision of the philosophical canon. The series also reveals the emancipatory potential inherent in the interplay between philosophy and other disciplines such as technology or aesthetics.

**Anna Elisabeth Höhl**, born in 1992, is a philosopher of science and works as a project coordinator at the Institute of Philosophy at Leibniz Universität Hannover. After degrees in physics, history and philosophy of science, she did her doctorate at Universität Bielefeld as a member of the DFG-funded Research Training Group "Integrating Ethics and Epistemology of Scientific Research". Her research focusses on topics in general philosophy of science and she approaches abstract questions through detailed analyses of concrete episodes from scientific practice.

Anna Elisabeth Höhl

# Scientific Understanding

What It Is and How It Is Achieved

[transcript]

This work was accepted as doctoral dissertation by the Faculty of History, Philosophy and Theology of Bielefeld University in 2022.
Examiners: Prof. Dr. Marie I. Kaiser, Prof. Dr. Mathias Frisch

This research was funded by the Deutsche Forschungsgemeinschaft (DFG, German Research Foundation) – Project 254954344/GRK2073.
This publication was funded by the Open Access Publishing Fund of the Leibniz Universität Hannnover and the Deutsche Forschungsgemeinschaft (DFG, German Research Foundation) – Project 254954344/GRK2073.

**Bibliographic information published by the Deutsche Nationalbibliothek**
The Deutsche Nationalbibliothek lists this publication in the Deutsche Nationalbibliografie; detailed bibliographic data are available in the Internet at https://dnb.dnb.de/



**First published in 2024 by transcript Verlag, Bielefeld**


Cover layout: Maria Arndt, Bielefeld
Printed by: Majuskel Medienproduktion GmbH, Wetzlar
https://doi.org/10.14361/9783839472620
Print-ISBN: 978-3-8376-7262-6
PDF-ISBN: 978-3-8394-7262-0
ISSN of series: 2941-8151
eISSN of series: 2941-816X

Printed on permanent acid-free text paper.

# Contents

**Acknowledgements** .......... 7

**List of Figures** .......... 9

**1. Introduction** .......... 11
1.1 Tracing understanding through the history of philosophy .......... 12
1.2 The neglect and (re-) discovery of understanding ... .......... 14
1.3 Outline of this book .......... 22

**2. Different views on scientific understanding** .......... 29
2.1 Henk de Regt: Understanding phenomena through theories .......... 31
2.2 Kareem Khalifa: Scientific understanding is scientific knowledge of an explanation.... 36
2.3 Finnur Dellsén: Understanding as dependency modelling .......... 39
2.4 Two questions concerning scientific understanding .......... 42

**3. Scientific understanding, scientific explanation, and why they cannot be torn apart** .......... 45
3.1 A few words on explanation .......... 46
3.2 Cases of understanding without explanation? .......... 49
3.3 Objectual and explanatory understanding – a controversy .......... 66
3.4 Further attempts to differentiate objectual and explanatory understanding .......... 76
3.5 Why scientific understanding and scientific explanation cannot be torn apart .......... 87

**4. Is scientific understanding an ability?** .......... 91
4.1 What are abilities? .......... 92
4.2 Understanding, an ability in itself .......... 111
4.3 The manifestation of scientific understanding .......... 124
4.4 The inextricable relation between understanding and knowledge .......... 136

**5. Scientific understanding of the genetic regulation of vertebrate development and how zebrafish made it possible** ..... 139

5.1 How zebrafish became a model organism: the integration of molecular genetics and developmental biology ..... 142

5.2 How is this a case of scientific understanding? ..... 157

5.3 Understanding the genetic regulation of vertebrate development scientifically ..... 169

**6. Grasping and explaining – an account of scientific understanding** ..... 173

6.1 The GE-account of scientific understanding ..... 173

6.2 Benefits of the GE-account of scientific understanding ..... 204

6.3 Understanding is an impressive cognitive achievement and a goal of science ..... 230

**7. Conclusion** ..... 233

7.1 Summary ..... 233

7.2 Outlook ..... 238

**Bibliography** ..... 249

# Acknowledgements

This book is the result of a doctoral project I have been conducting for three and a half years. Such complex and long-term projects are never realized or completed by one individual alone without the support and advice from other people. I had the good fortune to have had many excellent and fantastic people accompanying me throughout my time as a PhD student. I would like to take this opportunity to express my gratitude.

To begin with, I would like to thank my first supervisor, Prof. Dr. Marie I. Kaiser, for her support and guidance over the whole period of my doctoral project. She helped me to find and follow my way, already at the very beginning of the project, and at every point in time where I had to make crucial decisions. I very much appreciated her criticism, which always challenged me to scrutinize my thoughts and intuitions. Many thanks go to my second supervisor, Prof. Dr. Mathias Frisch, as well. He brought the topic of scientific understanding to my attention in the first place and gave me valuable feedback especially towards the end of the project.

Also, the wider environment in which I conducted my doctoral project contributed significantly to its successful completion. In particular, I would like to thank all the members of the Research Training Group 2073 in Bielefeld as well as in Hannover. Furthermore, I am grateful to the members of the Philosophy Department at Bielefeld University for creating and maintaining such a warm and friendly working atmosphere. Notably, my gratitude goes to all the members of the PhilBio Team and the PoS Seminar for reading and commenting on drafts of my work throughout the years. Additionally, I would like to thank Marcus Bachmann, Daphne Broeks, Linda Holland, Henk de Regt, Stefan Sleeuw and Javier Suárez for inspiring and in-depth discussion about the various dimensions and aspects of understanding.

Especially, my gratitude goes to Minea Gartzlaff, Ina Gawel, Saana Jukola, Milla Lifke, Rui Só Maia, Basel Myhub, Rose Trappes and Roel Visser for their assistance and friendship over the years. All of you made the time as a doctoral student much more enjoyable and an invaluable experience.

Finally, I would like to thank my family, without whose continuous support over the years I would not have made it to where I am now.

# List of Figures


Figure 1: An individual scientist S understands an empirical phenomenon P in a context C . 177
Figure 2: The manifestation of scientific understanding. ............................... 188
Figure 3: The resources for scientific understanding. ...................................197
Figure 4: The function of the scientific community for the understanding of S........... 203
Figure 5: The relations of understanding, explanation and prediction.................... 245

# 1. Introduction

Human beings strive to understand the world they inhabit. Aiming and attempting to understand something is ubiquitous in our everyday lives. Every human has witnessed situations in which, for example, a child wants to understand why one has to pay money when buying groceries from the supermarket, why someone is sad, or why her parents stay awake longer in the evening while she has to go to bed already. The same is of course also true of adults, although they may want to understand different things than children. Adult humans want to understand why a colleague is in a very bad mood, why there was a financial crisis, or how it was possible that populist parties gained more and more influence. And scientists strive to understand phenomena in the natural or social domains they are researching. Understanding phenomena is viewed to be one aim of science, as scientists occasionally state themselves. In biology, for instance, "a model organism [that] is a non-human species is extensively studied to *understand* specific biological phenomena."[1]

Despite the pervasive presence of instances and attempts of human beings to understand something in the world, the concept of *understanding*, what understanding is and how it is actually achieved, is hard to explicate. The uncertainty and confusion concerning the concrete meaning of *understanding* is not bound to a specific domain. From personal conversations with educators, I know that one goal of educational science is to develop tools or methods to determine whether pupils have understood what they are supposed to learn and understand in school. However, educators do this without having any clear concept or notion of *understanding*. They want to be able to measure something of which they have no idea what it actually is.[2] The controversial nature of understanding also becomes apparent in such provoca-

1 Sakaguchi, K. et. al. (2019), "Comprehensive Experimental System for a Promising Model Organism Candidate for Marine Teleosts." *Scientific Reports*, 9 (4948), DOI: 10.1038/s41598-019-41468-8, my emphasis.

2 This is the personal assessment of a friend of mine working in education science. Other educators might have a different opinion on that matter. However, if a fixed conception of understanding were employed in education science, my friend probably would not have a problem with it.

tive claims as "I think I can safely say that nobody understands quantum mechanics"[3] famously posed by Richard Feynman.

This book is a contribution to the philosophical research on (scientific) understanding. Surprisingly, philosophy has not paid much attention to understanding throughout its history, although understanding seems to be an ubiquitous epistemic activity as well as an unclear and contested concept. A philosophical interest in understanding emerged only quite recently within the last 25 years. Why has understanding been neglected by philosophers for such a long time and why has it attracted attention in recent times?

## 1.1 Tracing understanding through the history of philosophy

Understanding was already a topic of interest in ancient philosophy. In fact, it is proposed by some contemporary scholars that the more appropriate translation for the ancient Greek word *epistēmē* may be understanding, and not knowledge. For example, Julia Annas argues that in Plato's view, a person who has *epistēmē* does not merely possess various truths, but rather a systematic understanding of things.[4] Jonathan Lear makes a similar claim for Aristotle, namely that "to have *epistēmē* one must not only know a thing, one must also grasp its cause or explanation. This is to understand it: to know in a deep sense what it is and how it has come to be."[5] These interpretations of *epistēmē* reflect, according to Stephen Grimm, the currently widely accepted view that knowledge is quite easy to gain, while it seems harder to achieve understanding.[6] For example, I can know that I have blond hair just by looking at it, but understanding why I have blond hair or how I came to have blond hair demands something in addition. A reason for this difference might be that pieces of knowledge can be isolated or atomistic, at least in principle. I can know that I have blond hair without knowing anything else about, for example, genetics, inheritance, and the hair colors of my parents and ancestors. By contrast, targets of understanding seem to be more structured and interconnected. Although this might be due to the complex nature of targets that we want to understand, like the Second World War or the evolution of species, a certain degree of interconnectedness is also present in the understanding of isolated events. Understanding why a glass shatters requires connecting it with other events, for example the bumping of my elbow. Admittedly,

3 Feynman, R. P. (2017 [1965]), *The Character of Physical Law*. Cambridge (MA), MIT Press, p. 129.

4 See Annas, J. (1981), *An Introduction to Plato's Republic*. Oxford, Clarendon Press, chapter 10.

5 Lear, J. (1988), *Aristotle: The Desire to Understand*. New York, Cambridge University Press, DOI: 10.1017/CBO9780511570612.002, p. 6.

6 See Grimm, S., "Understanding", *The Stanford Encyclopedia of Philosophy* (Summer 2021 Edition), Edward N. Zalta (ed.), forthcoming URL = https://plato.stanford.edu/archives/sum2021/entries/understanding/ (last accessed April 11th, 2022), section 1.1.

the Greek philosophers seem to have held a very demanding concept of understanding that exceeds such mundane cases of understanding, as the understanding of the shattering of a glass, by far. Nevertheless, *epistēmē* is more similar to our contemporary notion of *understanding* than to the contemporary concept of *knowledge*, due to its emphasis on systematicity and interconnectedness.[7]

So, *understanding* is by no means a new concept in philosophy. Still, it has very much disappeared from the philosophical stage, and epistemologists focused instead on propositional knowledge. Why did that happen? No one really knows. Grimm presents two hypotheses. Perhaps the shift from understanding to knowledge was a reaction to the rise of skepticism in Hellenistic philosophy.[8] Alternatively or additionally, the wars of religion in 16th and 17th century Europe necessitated a focus on knowledge, as it became important to differentiate between good and bad knowledge claims.[9] Despite these first attempts at explaining the lack of understanding in the field of philosophy, the disappearance of understanding from philosophical debates is an unexplored issue so far, and I hope that research in history of philosophy will shed more light on it in the future.

It was not until the 19th century that the notion of understanding gained attention again in philosophy. Johann Gustav Droysen and Wilhelm Dilthey (re-)introduced understanding together with the *Verstehen-Erklären* dichotomy. It was their goal to elucidate the difference between the humanities (*Geisteswissenschaften*) and the natural sciences (*Naturwissenschaften*), and the *Verstehen-Erklären* dichotomy was taken to fulfill this purpose. For Droysen and Dilthey, understanding is the goal of the humanities, as the humanities are concerned with the intentions of (historical) actors and the interpretation of artefacts like texts or works of art. Therefore, understanding is subjective, according to Droysen and Dilthey. The natural sciences, in contrast, aim at uncovering the causes and general laws that are the basis of observed natural phenomena. Finding explanations of natural phenomena was, for Droysen and Dilthey, a purely objective endeavor.[10] Both scholars aimed at constituting an adequate theoretical as well as methodological basis for the "human sci-

---

7 See ibid.

8 See Zagzebski, L. (2001), "Recovering Understanding." In Steup, M. (ed.), *Knowledge, Truth, and Duty: Essays on Epistemic Justification, Responsibility, and Virtue*, pp. 235–252, New York, Oxford University Press, DOI: 10.1093/0195128923.003.0015.

9 For a systematic reconstruction of the history of epistemology, see Pasnau, R. (2017), *After Certainty: A History of Our Epistemic Ideals and Illusions*. New York, Oxford University Press, DOI: 10.1093/oso/9780198801788.001.0001.

10 See Beiser, F. C. (2011), *The German Historicist Tradition*. New York, Oxford University Press, DOI: 10.1093/acprof:oso/9780199691555.001.0001, chapter seven and eight; and Makkreel, R., "Wilhelm Dilthey", *The Stanford Encyclopedia of Philosophy* (Spring 2021 Edition), Edward N. Zalta (ed.), URL = https://plato.stanford.edu/archives/spr2021/entries/dilthey/ (last accessed April 11th, 2022).

ences", covering the humanities, to establish them as distinct from, but equally "scientific" as the natural sciences.[11] The *Verstehen-Erklären* dichotomy had a strong and complex influence on the Vienna Circle, and therefore on logical empiricism in general, and would maintain a long lasting impact on the philosophy of science.[12] However, apart from its prominent status in the debate about the *Verstehen-Erklären* dichotomy in the 19$^{th}$ century, understanding has never been seen as or taken to be an interesting or important topic for philosophers since antiquity. No one really cared about understanding.

## 1.2 The neglect and (re-) discovery of understanding ...

Given the absence of the notion of understanding from philosophical controversies for millennia, why has it become a topic for philosophy within the last few decades? Why do philosophers nowadays care about understanding? Baumberger, Beisbart & Brun identify three reasons for this trend:

> Understanding seems to be a central good that we try to realize when we think about the world. More specifically, the value of understanding seems to surpass that of knowledge. We can know something without understanding it. [...] The second reason for devoting attention to understanding is that understanding is a central goal of science. String theorist Greene goes so far as to characterize science in terms of understanding: "Science is the process that takes us from confusion to understanding in a manner that's precise, predictive and reliable." [...] The third reason to look at understanding derives from developments within epistemology.[13]

---

11 See for example Meinefeld, W. (1995), *Realität und Konstruktion. Erkenntnistheoretische Grundlagen einer Methodologie der empirischen Sozialforschung*, Wiesbaden, VS Verlag für Sozialwissenschaften, pp. 31–35, DOI: 10.1007/978-3-663-11243-3.

12 For more details on this relation, see for example Apel, K.-O. (1982), "The Erklären-Verstehen controversy in the philosophy of the natural and human sciences." In Fløistad, G. (ed.), *La philosophie contemporaine / Contemporary philosophy*, International Institute of Philosophy / Institut International de Philosophie, vol 2, pp. 19–49, Dordrecht, Springer, DOI: 10.1007/978-94-010-9940-0_2; or Uebel, T. (2010), "Opposition to *Verstehen* in Orthodox Logical Empiricism." In Feest, U. (ed.), *Historical Perspectives on Erklären and Verstehen*, pp. 291–309, Dordrecht, Springer, DOI: 10.1007/978-90-481-3540-0_15.

13 Baumberger, C., Beisbart, C. & Brun, G. (2017), "What is Understanding? An Overview of Recent Debates in Epistemology and Philosophy of Science." In Grimm, S., Baumberger C. & Ammon, S. (eds.), *Explaining Understanding. New Perspectives from Epistemology and Philosophy of Science*, pp. 1–34, New York and London, Routledge, pp. 2f; and Greene, B. (2008), "Put a Little Science in Your Life." New York Times, Open Ed., June 1, https://www.nytimes.com/2008/06/01/opinion/01greene.html (last accessed October 3rd, 2023).

Concerning the third reasons, the authors are referring to arising difficulties for and criticisms of justification conditions in accounts of knowledge. As these three reasons indicate, the two philosophical disciplines that are primarily concerned with understanding these days are philosophy of science and epistemology. Let us take a closer look at why understanding has not been worthy of inquiry for these disciplines for a long time, and why and how that changed.

### 1.2.1 ... in philosophy of science

The impact from the *Verstehen-Erklären* dichotomy, in which understanding was viewed as something subjective while explanation as something objective, became most apparent and was even amplified through the work of Carl Gustav Hempel in the 1960s. Due to his influential work, the notion of understanding was actively downplayed in philosophy of science for a long time. Essentially, Hempel was worried that understanding would be a threat to the objectivity of science. In Hempel's days, understanding was viewed as a subjective and psychological concept that – although it cannot be eliminated from science as science is conducted by human beings – should not be taken as a constitutive component of science. Explanation, on the contrary, is essential for science in Hempel's view. In a nutshell, the mere fact that some individual scientist does not understand some explanation of a phenomenon says nothing at all about the objectivity of the explanation. A good and objective scientific explanation of a phenomenon should be correct independently of any specific audience or context. As Hempel himself puts it:

> For scientific research seeks to account for empirical phenomena by means of laws and theories which are objective in the sense that their empirical implications and their evidential support are independent of what particular individuals happen to test or to apply them; and the explanations, as well as the predictions, based upon such laws and theories are meant to be objective in an analogous sense. This ideal intent suggests the problem of constructing a non-pragmatic concept of scientific explanation.[14]

Only few philosophers offered resistance to Hempel's subjectivist view of understanding, but there were some. As early as 1974, Michael Friedman argued that just because understanding seems to have some psychological element, it does not follow that understanding is purely subjective, uninteresting or unimportant for philosophers of science.[15] Eventually, knowledge has a psychological element, too, because

14 Hempel, C. G. (1965), *Aspects of Scientific Explanation and Other Essays in the Philosophy of Science*. New York, Free Press, p. 426.

15 See Friedman, M. (1974), "Explanation and Scientific Understanding." *Journal of Philosophy*, 71 (1), pp. 5–19, DOI: 10.2307/2024924.

of the belief condition, but this does not amount to a degradation of knowledge as being purely or mainly subjective. And Jaegwon Kim pointed out that humans aim at explaining things because we want to understand them. Hence, in Kim's view, it is a grave mistake to separate accounts of explanation from understanding.[16]

However, proponents of understanding could not remove the reservations about this notion in philosophy of science for a significant period of time. Even in the early 2000s, suspicions concerning understanding were actively endorsed, for example by John Trout, who defended the Hempelian objectivist account of explanation as well as his subjectivist view of understanding. According to Trout, understanding merely is a *Eureka!* or *Aha!* feeling, a product of types of biases known in cognitive psychology: "This sense of understanding alone is not necessarily a reliable guide to truth, nor is it a necessary condition for good explanation. Still less is it sufficient for good explanation."[17] A significant change of this situation was only brought about by the distinction between the "feeling" or phenomenology of understanding and genuine understanding introduced by Henk de Regt.[18] He also extensively engaged with Hempel's work in order to argue that understanding should not be taken as purely subjective and pragmatic. De Regt should be successful. By now, understanding is receiving more and more attention within philosophy of science, it is viewed as a legitimate and interesting topic for philosophers of science, and de Regt can be seen as the most important founding father of the contemporary debate on understanding in this discipline.

While understanding is a respected topic within philosophy of science by now, specific questions or issues concerning understanding receive special attention. According to Grimm, philosophers of science are particularly interested in the relation of understanding and explanation, as well as the relation of understanding and idealization, where the notion of idealization is intended to cover idealized models and representations.[19] Concerning the first issue, "the relationship between explanation and understanding remains controversial."[20] Grimm differentiates between two main approaches to this topic, which are the "understanding-first" approach versus the "explanation-first" approach. The "understanding-first" approach takes understanding to be "conceptually prior to, or more basic then, the notion of expla-

---

16 See Kim, J. (1994 [2010]), "Explanatory Knowledge and Metaphysical Dependence." *Philosophical Issues*, 5, pp. 51–69, Reprinted in Kim, J. (2010), *Essays in the Metaphysics of Mind*. New York, Oxford University Press, DOI: 10.1093/acprof:oso/9780199585878.001.0001.

17 Trout, J. D. (2002), "Scientific Explanation And The Sense Of Understanding." *Philosophy of Science*, 69 (2), pp. 212–233, DOI: 10.1086/341050, p. 213.

18 See de Regt, H. W. (2004), "Discussion Note: Making Sense of Understanding." *Philosophy of Science*, 71 (1), pp. 98–109, DOI: 10.1086/381415.

19 See Grimm (2021), section 4.

20 Ibid.

nation."[21] That is, explanation serves the end of understanding, and the goodness or "explanatoriness" of any explanation is assessed in terms of its capacity to generate understanding. Proponents of the "understanding-first" approach include Daniel Wilkenfeld, Paul Humphreys, and Angela Potochnik, among others.[22] By contrast, according to the "explanation-first" approach, most prominently endorsed by Kareem Khalifa, the discussion of understanding is nothing but a repacking of existing theories of explanation:

> All one needs for a plausible account of understanding is a plausible account of what counts as a good or correct explanation, combined with a plausible account of knowledge. Understanding therefore amounts to *knowing a correct explanation*. But then nothing new or special is needed to theorize about understanding; our accounts of explanation and our theories of knowledge do all the important theoretical work.[23]

Intuitions regarding these two approaches vary significantly, as the basic issue concerns the question of whether or not understanding *why p* does in some way exceed knowing *why p*.

The second interesting issue for philosophers of science, according to Grimm, is the relation of understanding and idealizations. Like explanation, idealization is a long-standing central and established topic in philosophy of science. Explanation as well as idealization are ubiquitous in science. The central challenge posed by idealization is to account for their capacity to provide real epistemic benefits without adhering to the truth (at least not in a strict or straightforward sense). All idealizations misrepresent or falsify the world in some way, e.g. by appealing to fully rational agents or frictionless planes, or through omitting certain factors, e.g. long range intermolecular forces.[24]

> Yet if idealizations provide epistemic benefits, and we cannot readily think of the benefits in terms of truth, then how exactly should we think about them? According to some philosophers, we should think not in terms of truth but rather in terms

---

21 Ibid.

22 For more details, see e.g. Wilkenfeld, D. A. (2013), "Understanding as Representation Manipulability." *Synthese*, 190 (6), pp. 997–1016, DOI: 10.1007/s11229-011-0055-x; and Humphreys, P. (2000), "Analytic Versus Synthetic Understanding." In Fetzer, J. (ed.), *Science, Explanation, and Rationality: The Philosophy of Carl G. Hempel*, pp. 267–286, Oxford, Oxford University Press, DOI: 10.1093/oso/9780199334872.003.0017; and Potochnik, A. (2017), *Idealization and the Aims of Science*. Chicago (IL), University of Chicago Press, DOI: 10.7208/9780226507194.

23 Grimm (2021), section 4.1; see especially Khalifa, K. (2017b), *Understanding, Explanation, and Scientific Knowledge*. Cambridge, Cambridge University Press, DOI: 10.1017/9781108164276.

24 See Grimm (2021).

> of understanding. Understanding is the epistemic benefit we receive from idealizations, and understanding and truth can come apart. On this view, understanding (unlike knowledge) can therefore be "non-factive".[25]

Concerning the details of how exactly idealizations might provide understanding, different accounts have been offered, for example by Angela Potochnik, Catherine Elgin, and Michael Strevens.[26] Furthermore, there is no general consensus that understanding can be non-factive. To put it differently, it is highly contested whether a representation that does not (somehow) answer to the facts can enable (genuine) understanding.[27]

Thus, the core topics surrounding understanding (although not the only ones) that philosophers of science are interested in relate to two traditional, central issues within the field, explanation and idealization. However, the second philosophical discipline that has also (re-)discovered understanding as a topic of interest is epistemology. Why did epistemologists become interested in understanding?

### 1.2.2 ... and in epistemology

In epistemology, the notion of understanding has not been actively devalued as in the philosophy of science, but rather neglected or ignored, according to Grimm.[28] As stated in section 1.1, the reasons for this disinterest in understanding throughout the history of western philosophy are not really known, but it is possible to identify reasons why this situation changed, i.e. why epistemologists started to be interested in understanding towards the end of the 20$^{th}$ century (again). Grimm identifies three reasons for this development. These three reasons are in line with, though not identical to, the reasons that Baumberger, Beisbart & Brun present for this trend, mentioned in section 1.2.

The first observation by Grimm is that, as especially Elgin argued, some of the greatest intellectual achievements of humanity, which can be found, for instance, in the sciences and arts, are actually not targeted at what epistemologists traditionally view as knowledge, but rather at something that is more plausibly conceptualized as

25 Ibid, section 4.2.

26 See Potochnik (2017), Elgin, C. Z. (2017), *True Enough*. Cambridge (MA), MIT Press; and Strevens, M. (2017), "How Idealizations Provide Understanding." In Grimm, S., Baumberger C. & Ammon, S. (eds.), *Explaining Understanding. New Perspectives from Epistemology and Philosophy of Science*, pp. 37–49, New York and London, Routledge.

27 For an overview of different positions and arguments in favor of and against (moderate) factivity and non-factivity, see Baumberger, Beisbart, & Brun (2017).

28 See Grimm (2021), section 1.2.

understanding.[29] Second, he argues that the interest in understanding was driven by virtue epistemologists. For example, Linda Zagzebski maintained that the focus of epistemology on knowledge as one epistemic good is too narrow, as it does not do justice to other highly valued epistemic goods like e.g. wisdom or understanding.[30] And third, Jonathan Kvanvig influentially argued that understanding is distinctively valuable, while knowledge is not.[31] Grimm concludes that all the different lines of arguments accused the field of epistemology of having a too narrow and one-sided focus on knowledge and that "epistemology needed to be broadened so that goods such as understanding could be given their proper due, and their claims resonated with other epistemologists."[32]

The first observation presented by Grimm supports the second reason for the new interest in understanding presented by Baumberger, Beisbart & Brun, namely that understanding is a central goal of the sciences (and potentially also of other areas of human action). Grimm's second and third observation concerning the views from (virtue) epistemology resembles the first reason provided by Baumberger, Beisbart & Brun, which is the (potential) special epistemic value of understanding. Whatever understanding is, it seems to be something that goes over and above knowledge (and the same could be argued for wisdom, for example). If this is the case, then understanding is epistemically more valuable than knowledge, and the traditional focus on knowledge in epistemology becomes questionable.

> [In this context, it has been noted several times that] knowledge may easily be acquired through the testimony of experts; understanding, by contrast, seems more demanding and requires that an epistemic agent herself puts together several pieces of information, grasps connections, can reason about causes, and this too suggests an added value. [...] The problem of accounting for a supposed special value of knowledge is now called *the value problem for knowledge*. Epistemology escapes this problem if it turns to understanding.[33]

---

29 See particularly Elgin, C. Z. (1991), "Understanding: Art and Science." *Midwest Studies in Philosophy*, 16, pp. 196–208, DOI: 10.1111/j.1475-4975.1991.tb00239.x. For a more recent elaboration of her ideas concerning understanding, see Elgin (2017).

30 See Zagzebski, L. (1996), *Virtues of the Mind. An Inquiry into the Nature of Virtue and the Ethical Foundations of Knowledge*, New York, Cambridge University Press, DOI: 10.1017/CBO9781139174763; and Zagzebski (2001).

31 See Kvanvig, J. L. (2003), *The Value of Knowledge and the Pursuit of Understanding*. New York, Cambridge University Press, DOI: 10.1017/CBO9780511498909.

32 Grimm (2021), section 1.2.

33 Baumberger, Beisbart & Brun (2017), p. 3, original emphasis. For a detailed discussion of *the value problem for knowledge*, see Pritchard, D., Millar, A. & Haddock, A. (2010), *The Nature and Value of Knowledge: Three Investigations*, Oxford, Oxford University Press, DOI: 10.1093/acprof:oso/9780199586264.001.0001. Jonathan Kvanvig also explicity advocates the special value of understanding in comparison to knowledge, see for example Kvanvig (2003).

However, and quite unsurprisingly, the alleged special epistemic value of understanding is not universally accepted and objections are put forward.[34]

Investigations on and discussions about the possible special epistemic value of understanding are related and intertwined with the third reason that Baumberger, Beisbart & Brun identify for the interest in understanding, namely other developments within epistemology that concern the notion of justification. Apparently, strong intuitions persist that a justification for a belief, which would turn the belief into knowledge, is on the one hand accessible for the epistemic agent, is internal to her, and on the other hand that the belief is justified through being embedded in a coherent web of beliefs. These views on justification for beliefs are called internalism and coherentism.[35]

> [Baumberger, Beisbart & Brun observe that while] intuitions supporting internalism and coherentism seem deep-seated, it has been proven difficult to save them in an account of knowledge. Internalism about epistemic justification is threatened by a regress problem. Coherentists have a hard time to show how coherence is related to truth, which is supposed to be the aim of belief and a central feature of knowledge. However, an immediate access to the reasons for a belief and the ability to connect a belief with others seem to be central for understanding.[36]

Hence, they conclude that internalist and coherentist intuitions can appropriately account for understanding, while they are (or might be) inapplicable to knowledge. Again, such views that allocate internalist or coherentist intuitions to understanding, and additionally intuitions from virtue epistemology, are contested as well.[37]

---

34 See for instance Carter, J. & Gordon, E. (2014), "Objectual Understanding and the Value Problem." *American Philosophical Quarterly*, 51 (1), pp. 1–13; or Khalifa (2017b), chapter 8.

35 See for example Pappas, G., "Internalist vs. Externalist Conceptions of Epistemic Justification", *The Stanford Encyclopedia of Philosophy* (Fall 2017 Edition), Edward N. Zalta (ed.), URL = https://plato.stanford.edu/archives/fall2017/entries/justep-intext/ (last accessed April 11th, 2022); and Olsson, E., "Coherentist Theories of Epistemic Justification", *The Stanford Encyclopedia of Philosophy* (Fall 2021 Edition), Edward N. Zalta (ed.), URL = https://plato.stanford.edu/archives/fall2021/entries/justep-coherence/ (last accessed April 11th, 2022).

36 Baumberger, Beisbart & Brun (2017) pp. 3f.

37 See for instance Grimm, S. (2017), „Understanding and Transparency." In Grimm, S., Baumberger C. & Ammon, S. (eds.), *Explaining Understanding. New Perspectives from Epistemology and Philosophy of Science* pp. 212–229, New York and London, Routledge; and Khalifa, K. (2017a), „Must Understanding be Coherent?" In Grimm, S., Baumberger, C. & Ammon, S. (eds.), *Explaining Understanding. New Perspectives from Epistemology and Philosophy of Science*, pp. 139–164, New York and London, Routledge.

### 1.2.3 The current state of play

Where does all this leave us? In sum, various reasons have led to the recent interest of philosophers in understanding. Most fundamentally, it has been acknowledged that understanding is a central epistemic good that humans try to and want to achieve, in science as well as elsewhere. Because of the recognition of this central value of understanding, philosophers of science and epistemologists began paying attention to the concept. The engagement with understanding has opened up new possibilities to solve already known problems and answer open questions in the respective disciplines. For philosophy of science, the examination of understanding enables analyses of a crucial (or potentially the most important) scientific aim and the function of idealizations in scientific research, among others. In epistemology, the work on understanding offers solutions to central issues in the field, such as *the value problem for knowledge*, or problems concerning different accounts of epistemic justification.

Furthermore, the topic of understanding has yielded an extensive exchange between these two philosophical disciplines, which is a rare phenomenon. Philosophers of science and epistemologists engage with and criticize each other's work in this area. This interdisciplinary research on the topic of understanding has resulted in a fruitful dialogue between the two philosophical disciplines, each of which contributes their own main interests, questions, and respective methods. Philosophy of science is interested in the understanding that scientists achieve of the phenomena they are researching or of the theories that they use, how scientists achieve this understanding, and how scientific understanding relates to other core concepts in science, such as theories, explanations, and models, among others. Epistemology, in contrast, is interested in the understanding that (human or rational) agents generally can gain.

Despite their differences, philosophers of science and epistemologists share the common ground in that the concept of understanding should be analyzed and provide each other with important insights. On the one hand, epistemologists (usually) take scientific understanding to be a special or in some sense distinctively valuable type of understanding, and hence they are interested in what philosophers of science have to say about it. On the other hand, any general inquiry into understanding in epistemology might reveal important insights for philosophers of science.

This dialogue between the two philosophical disciplines amounts to ever new research questions on understanding, including the following:

- What is understanding? Is it a kind of knowledge or an ability?
- What types of understanding exist and how might they relate to each other (e.g. objectual, explanatory, symbolic, practical, moral understanding …)?
- Does understanding require explanation?
- How does understanding relate to truth?

- What is the role of models or representations in or for understanding?
- How should the graduality of understanding be accommodated?
- ...

This list is not meant to be exhaustive, it shall just exemplify the diversity and amount of questions concerning understanding which puzzle philosophers.[38] This book is a contribution to this thriving and new 'hot' topic and debate in philosophy.

## 1.3 Outline of this book

The goal of this book is to provide a philosophical account of scientific understanding. That is, I develop and defend necessary and sufficient conditions for the understanding that scientists achieve of the phenomena they are studying through their research. To put it differently, I provide answers to these two main questions:

I) What is scientific understanding?
II) And how do scientists achieve understanding?

My starting point is the existing philosophical debate on understanding, including work on scientific understanding in particular, as well as on understanding in general. As I have stressed throughout the previous section, understanding is a topic of interest for philosophers of science as well as for epistemologists for the last three decades. Hence, perspectives and insights from both disciplines need to be considered in any analysis on understanding. Chapters two, three and four cover, therefore, abstract and normative philosophical discussions and positions concerning characteristics of understanding and related concepts, such as explanation and ability or knowing-how. The philosophical literature already has a lot to offer for clarifying the notion of understanding, and I could only scratch the surface of the variety of topics and issues around understanding that are addressed in philosophy by now in the previous section.

Engaging with different and opposing views and arguments allows for the identification of core topics concerning understanding, together with underlying assumptions or intuitions regarding certain issues. A thorough examination of the existing philosophical literature on understanding provides an important navigation through the debate. It enables structuring the debate into more or less controversial or important topics with regard to specific questions, together with the genesis of certain camps within the debate, the arguments developed by proponents of

38 Further summaries of the philosophical debate on understanding and the questions that occupy philosophers are provided by Baumberger, Beisbart & Brun (2017) and Grimm (2021).

different views, and also the intuitions that are uphold within and across different camps. Taking already developed philosophical arguments and theories about aspects of understanding enables me to clarify my intuitions concerning understanding, why I agree with some scholars and not with others, and developing my own normative position on questions such as why scientific understanding should require explanation or why understanding should be conceptualized as an ability.

However, while abstract philosophical theorizing and argumentation is of course important, it is not sufficient for generating a meaningful account of scientific understanding. I am a philosopher of science and, hence, have acquired the specific methods and practices of my discipline. Since the Practice Turn in 1980s, through which philosophical as well as historical and social studies of science paid much more attention to the social, psychological, and material dimensions of scientific research, philosophy of science is (often) characterized by in-depth and detailed case studies or episodes from past or present scientific practice. I adhere to this practice, as it is my deepest conviction that philosophy has to take scientific practice seriously if it wants to generate and contribute any relevant or important insights about science.[39] The most consistent and intuitive philosophical theory or argument will be meaningless if it cannot capture real and actual science.

Thus, while I draw on shorter examples from scientific as well as none-scientific contexts throughout my abstract philosophical discussions to support my claims, these examples cannot replace in-depth analyses of research episodes. Examples serve the purpose of illustrating or supporting claims or arguments. Analyses of research episodes shall, additionally, provide novel insights into the topic of interest, scientific understanding in this case. Hence, an analysis of an episode from scientific practice will follow my abstract philosophical discussions. Thereby, I can reassess the plausibility of my claims developed before, and also highlight novel insights about how scientists gain understanding that philosophical theorizing cannot deliver on its own. Taken together, the insights from my detailed engagement with the abstract philosophical debate and from the analysis of the concrete research episode allow me to generate a novel account of scientific understanding that will be meaningful for science in general.

Let me be more specific about the structure of this book. While insights from philosophy of science as well as epistemology need to be taken into account in any

39 For more information about the engagement of philosophy of science with scientific practice, see e.g. Soler, L., Zwart, S., Lynch, M. & Israël-Jost, V. (eds.) (2014), *Science After the Practice Turn in the Philosophy, History, and Social Studies of Science*. Studies in the Philosophy of Science 14, New York and London, Routledge. For insights into the relation between history and philosophy of science in particular, see Schmaltz, T. & Mauskopf, S. (eds.) (2012), *Integrating History and Philosophy of Science: Problems and Prospects*. Dordrecht, Springer, DOI: 10.1007/978-94-007-1745-9.

analysis on understanding, this does not prohibit placing an emphasis on one of the respective disciplines. Again, I am a philosopher of science by training and approach the topic from the corresponding perspective. Thus, I start by introducing already existing accounts of scientific understanding put forward by the three philosophers of science Henk de Regt, Kareem Khalifa and Finnur Dellsén in chapter two. Presenting their accounts provides the basis for my analysis and allows for identifying central issues, question, and also shared assumptions related to scientific understanding. I restrict myself to the accounts of scientific understanding developed by philosophers of science in chapter two because this book focusses on scientific understanding. While I hope that my work will also be of value for epistemologists, I do not aim at providing an account of understanding in general. Hence, while I give epistemology its earned space throughout this book, especially in chapters three and four, I do not introduce any epistemological account of understanding at the beginning of my investigations in chapter two. Instead, I highlight the agreements and disagreements among de Regt, Khalifa, and Dellsén concerning scientific understanding and argue that two questions are especially striking:

1) Does scientific understanding require explanation or not?
2) Is understanding an ability or a type of knowledge?

These two questions provide partial answers to one of the main questions of this book, namely what scientific understanding is. The answer to the second main question, how scientists can or do achieve understanding, depends, in turn, on what scientific understanding is. In order to clarify what scientific understanding is and how scientists achieve it, which is the ultimate goal of this book, I first need to figure out whether understanding requires explanation and whether it is an ability or a type of knowledge. Thus, this book is structured as follows:

Having provided a starting point in chapter two, chapter three turns towards the question of whether scientific understanding requires explanation or not, a core question for philosophers of science, as I have indicated in section 1.2.1. Answering this question is not possible without clarifying what I mean by explanation. A clarification of the notion of explanation will be the first step to take in this chapter. Following that, I engage with several arguments and examples put forward to defend a separation or possible independence of understanding from explanation and challenge their plausibility for scientific understanding in particular. One line of thought is that understanding and explanation can fall apart, as the former is something tacit, while the latter is something explicit. Is it really that simple, is this straightforwardly the case? Furthermore, the philosophical literature on understanding intensively engaged with two types of understand so far: explanatory and objectual understanding. Explanatory understanding is viewed to require explanation. Objectual understanding, in contrast, is either independent of explanation or exceeds explanatory understanding in some sense. How exactly do different scholars distinguish between these two types of understanding? And do the proposed distinctions really show a difference between two kinds of understanding, or do the differences merely point towards a difference in the degree of understanding? In other words, can objectual understanding be conceptualized as being independent of explanation? And finally, in light of the discussions conducted in this chapter, does it make sense to view scientific understanding as being possibly independent of explanation?

Chapter four addresses the question of whether understanding is an ability or a type of (propositional) knowledge. This chapter will probably be especially interesting for epistemologists, as it touches on central issues discussed in this field mentioned in section 1.2.2, including the demand for broadening epistemology's focus to other intellectual achievements than knowledge. Similarly to my approach in chapter three, I first need to clarify what I mean by ability. I will do that based on a broad literature survey covering work on abilities or knowing-how from philosophy of science, (virtue) epistemology, and metaphysics. Having generated a conceptualization of ability, the follow up question will be whether understanding should be viewed as an ability, or rather as a type of (propositional) knowledge. Can the notion of understanding and potential characteristics that are often intuitively ascribed to understanding be better accommodated by conceptualizing understanding as an ability, and not as a type of (propositional) knowledge? And if the answer to this question is affirmative, if understanding is taken to be an ability, how could understanding be manifested? In contrast to chapter three, where I mainly use and engage with the philosophical literature on understanding, my analysis in chapter four will be broader. Answering the questions of what abilities are, whether understanding should be conceived as an ability and if so, how understanding is manifested, requires a wider perspective that includes but also exceeds the debate on understanding, as more general questions concerning the nature of (propositional) knowledge and abilities need to be taken into account and then applied to discussions on understanding.

Chapter five then presents a change of perspective and differs significantly from the previous chapters, as it deals with an episode from scientific practice and the question how exactly scientists involved in this episode gained scientific understanding. The research episode covers the introduction of zebrafish as a new model organism into biological research through which biologists wanted to understand the genetic regulation of vertebrate development. A careful and detailed description of this research episode will, on the one hand, allow for testing whether the insights gained through abstract philosophical argumentation in chapters three and four have any meaningful implications for reals world science. On the other hand, analyzing the episode might allow for an identification of features of understanding that cannot be obtained by pure philosophizing detached from any real-world cases. What do real scientists need in order to understand a phenomenon they are researching? How do they actually achieve understanding? How do they proceed? Which steps do they undertake? This chapter aims at providing answers to these questions.

Finally, in chapter six, I bring all the different lines of thought developed in the previous chapters together and present my account of scientific understanding. This account will be an answer to the main questions of what scientific understanding is and how scientists achieve it, and will go beyond the concrete episode discussed in the previous chapter. Of course, the episode from biology serves an important purpose, but one cannot straightforwardly generalize from that episode to science as a whole, given the variety of different scientific disciplines. That is, the concrete insights from chapter five need to be abstracted in order to be relevant for scientific understanding, generally speaking, and these abstracted insights still need to be connected to the results from chapters three and four. How are understanding and explanation related? How do understanding, knowledge, and abilities tie in together? What do scientists need in order to understand a phenomenon? Which roles do the respective context and the research community in which any particular scientists is embedded play for understanding? In short, I will explain what I take scientific understanding to be and how scientists achieve it. To make my account of scientific understanding even more appealing, I will come back to the accounts by Henk de Regt, Kareem Khalifa and Finnur Dellsén introduced in chapter two, and I will emphasize the benefits of my account in comparison to their alternatives. Ultimately, chapter seven will provide a summary of the findings of this book and an outlook by pointing towards further questions and issues concerning scientific understanding that lie beyond the scope of this book.

In short, this book can be divided, roughly, into two parts. In chapters two, three, and four, I provide a detailed account of the philosophical debate on understanding, highlight common assumptions and intuitions as well as opposing positions and arguments, and I also consider philosophical work on related topics, in order to develop my arguments and position concerning the respective issue addressed in these chapters. Chapters five and six make up the second part of the book, as I answer the two main questions guiding my project, namely what scientific understanding is and how scientists achieve it. While my results from the first part of course contribute to the second part, my analysis and argumentation in chapters five and six are not as closely tied to the existing philosophical debate on understanding as the three preceding chapters.

As a general outcome, this book will amplify and further support recent developments within epistemology and philosophy of science as regards (scientific) understanding. My aim is to second the currently dominant view that understanding is an important intellectual achievement and goal of scientific research worthy of philosophical attention and analysis. Viewing understanding as something purely subjective or psychological that has no legitimate place in the natural sciences, as argued by Droysen and Dilthey in the $19^{th}$ century and by Hempel and other philosophers of science throughout the $20^{th}$ century, is unjustified in light of the importance attributed to understanding in scientific as well as everyday contexts. Valuing and researching understanding enables new perspectives and insights not only on scientific practice, but also the nature of knowledge and central epistemic goods more generally. But first things first, let us start by taking a look at some existing ideas and accounts of scientific understanding.

# 2. Different views on scientific understanding

As already mentioned in the introduction, understanding has attracted the attention of epistemologists and philosophers of science only within the last 25 to 30 years. Since then, more and more philosophers have engaged with the investigation on understanding, be it scientific understanding in particular, or understanding in general. As it can be expected, philosophers of science are primarily concerned with scientific understanding, whereas epistemologists turn to understanding as a general phenomenon or aim of human beings. However, these two philosophical disciplines talk to and with each other. Even epistemologists refer to scientific understanding quite often, as they take scientific understanding to be an especially valuable kind of understanding, maybe even the best kind of understanding one could achieve. The dialogue between philosophers of science and epistemologists is, hence, a very fruitful one. However, as this book focusses on scientific understanding, I will primarily rely on work done by philosophers of science, while I will also give epistemology its proper space in the course of my investigation.

In this chapter, I give an overview of three very important and influential accounts of scientific understanding. These are the accounts developed by Henk de Regt, Kareem Khalifa, and Finnur Dellsén. Based on this overview, I will highlight common assumptions shared by those three scholars, as well as disagreements among them concerning characteristics and the nature of understanding. The identification of commonalities of the three accounts will provide a basis for my own work on scientific understanding. And the detection of dissent and conflicting arguments will allow me to spotlight two questions that are of central importance for any analysis of scientific understanding.

Why do I only address these three accounts, although de Regt, Khalifa and Dellsén are by far not the only philosophers engaged in the discussions about understanding? While more and more epistemologists and philosophers of science join the debate about understanding, most of them focus on specific features of understanding, like its relation to truth, to idealizations or to explanations, to mention just some examples. To my knowledge, de Regt and Khalifa are the only scholars so far who developed full-fledged and detailed philosophical accounts of scientific understanding that they presented, defended and published in form of monographs. Since their accounts are the most elaborated ones on the market, they certainly need to be considered in any philosophical investigation on scientific understanding. Additionally, I also include the account of scientific understanding developed by Dellsén. The reason to add this third account to the discussion in this section is that de Regt and Khalifa, despite all their differences, agree in one central aspect: they take explanation to be necessary for understanding. Their accounts only accommodate what is often called explanatory understanding. However, as I show in section 2.3 and especially in chapter three, there is no universal agreement that understanding requires explanation. On the contrary, accounts of understanding without explanation, most often called objectual understanding, are developed and defended in the literature. Therefore, I need to take an account of objectual understanding into consideration, and the one developed by Dellsén is a good representative for such accounts. Taken together, the accounts of scientific understanding from de Regt, Khalifa and Dellsén sufficiently represent the range of views concerning features that most scholars in the debate take to be crucial for understanding: abilities in general, grasping in particular, knowledge and explanation.

I do not address any account of understanding developed by epistemologists in this section, because epistemologists are usually interested in accounts of understanding in general, not with scientific understanding in particular. So, epistemologists aim at accounts that should accommodate understanding that any human agent could gain, not only scientists. While I do think that it might be possible to develop a unified or holistic account of understanding and that it is reasonable to expect some commonalities of scientific understanding and, let's say, laypeople's understanding, it is also plausible to expect some differences between scientific and other types of understanding. It is not the goal of this book to develop a general account of understanding that human beings can gain of anything. I happily leave this task to epistemologists. Instead, I want to develop an account of scientific understanding, the understanding that scientists achieve of the phenomena they are researching. Hence, I focus and only introduce three accounts of scientific understanding put forward by philosophers of science in this chapter. However, I will let epistemologists engaged in understanding have their say, especially in chapters three and four, where I discuss the relation of understanding and explanation as well as the nature of understanding.

But first, let us have a look what other philosophers of science think about scientific understanding. I start with Hend de Regt's account in section 2.1, followed by Kareem Khalifa's account in section 2.2, and by Finnur Dellséns's account in section 2.3. Finally, in section 2.4, I highlight in which aspects concerning understanding those three scholars agree, and on which they disagree and in what way. By doing so, I will point out that two crucial questions arise, and that I am going to answer these two questions in the course of this book. So, let's get started.

## 2.1 Henk de Regt: Understanding phenomena through theories

Henk de Regt was among the very first philosophers of science who were interested in understanding as a topic worthy of analysis for its own sake. Hence, he is something like a founding father of the philosophical engagement with understanding in contemporary philosophy. Any work on scientific understanding should consider de Regt's contribution. In his analysis, de Regt differentiates between two kinds of understanding crucial in science, which he characterizes as follows:

> **UP:** understanding a phenomenon = having an adequate explanation of the phenomenon (relating the phenomenon to accepted items of knowledge)

> **UT:** understanding a theory = being able to use the theory (pragmatic understanding)[1]

Additionally, he differentiates between three levels of science: the macro-level (that refers to science as a whole), the meso-level (that characterizes scientific communities) and the micro-level (corresponding to individual scientists). UT is intended to allow for micro-and meso-level variations, which means that different explanatory strategies are applied to achieve the macro-level aim UP.[2]

At the core of de Regt's account of understanding lies the thesis that scientists need intelligible theories if they want to gain scientific understanding of phenomena. If a theory is not intelligible for scientists, they will not be able to use the theory to construct an explanation of a phenomenon on the basis of the theory. Without understanding a theory, understanding a phenomenon is impossible.[3] This leads de Regt to his criterion for understanding phenomena:

> **CUP:** A phenomenon P is understood scientifically if and only if there is an explanation of P that is based on an intelligible theory T and conforms to the basic epistemic values of empirical adequacy and internal consistency.[4]

Whether a theory is intelligible or not depends on the epistemic framework that is accepted and provided by a scientific community. The available background knowledge and acquired skills determine whether a theory is intelligible. In different disciplinary, historical or social contexts, different theories are judged as being intelligible by scientists. To restrain the scientist's freedom to choose or construct any

---

1 De Regt, H. W. (2017), *Understanding Scientific Understanding*. New York, Oxford University Press, DOI: 10.1093/oso/9780190652913.001.0001, p. 91.

2 See ibid. pp. 90f.

3 See ibid. pp. 91f.

4 Ibid. p. 92.

kind of explanation based on any theory taken to be intelligible, de Regt adds the two epistemic values of empirical adequacy and internal consistency. Although it is not always obvious how these values should be applied, some degree of adequacy and consistency has to be reached by any explanation to secure a minimal scientific standard.[5]

De Regt anticipates possible critique to CUP in the form of putting too much emphasize on theories in the process of achieving understanding. Sometimes, scientists achieve understanding merely by experimentation that is independent of any theory. Additionally, there are scientific disciplines where explicit theories are absent. Finally, models constructed and used in science seem to provide understanding independently of any theory. De Regt's answer to these objections is that he does not require any theory to be full-fledged or explicitly articulated. Instead, he follows Giere[6] and views a theory to be a collection of principles, which guide every form of observation, experiment, or model-construction. No scientific discipline works without some theoretical framework, no matter how explicit that might be.[7]

The account of scientific understanding developed by de Regt accommodates solely explanatory understanding, the understanding that is produced by a scientific explanation. Although he mentions that other forms of understanding exist, he does not cover or address them in his account. Rather, he simply presupposes their existence and characterizes them as types of knowledge. Since it is not possible in de Regt's framework to achieve explanatory understanding if there is no theory (no theoretical principles at all), theories are also viewed as a part of knowledge. De Regt employs a generic conception of explanation, which characterizes every explanation as an argument that presents a systematic line of reasoning that connects a phenomenon to accepted theoretical and empirical background knowledge. Hence, theories have to be part of the background knowledge. Otherwise, there will be nothing a phenomenon could be connected to, therefore there will be no explanation and, hence, also no explanatory understanding. An intelligible theoretical context is required for achieving explanatory understanding.[8]

Another argument that understanding and intelligibility are deeply intertwined is rooted in the fact that scientists understand the phenomena of their research by interacting and communicating with each other. If the theoretical principles, assumptions or models used by some scientists are not intelligible for others, science as a community will not make any progress. In most cases, scientists address and solve problems together in cooperation. Therefore, everything that contributes to the common activity of scientific research has to be intelligible for its members. De

5 See ibid. pp. 92f.

6 See Giere, R. N. (2006), *Scientific Perspectivism*. Chicago, University of Chicago Press, pp. 59–69.

7 See de Regt (2017), pp. 95ff.

8 See ibid. pp. 96–99.

Regt also describes the connection between his notion of intelligibility and the account of Humphreys[9], who introduces primary and secondary understanding. Primary understanding is achieved by an individual or group of scientists in isolation (would be identical to UT, having constructed an explanation), whereas secondary understanding requires communication and making the new constructed explanation intelligible to other individuals.[10]

Since achieving scientific understanding depends on intelligible theories, de Regt has to determine under which conditions a theory is intelligible. He wants to set some restrictions so that scientists do not decide on a purely subjective basis which theory is intelligible under which circumstances. If a theory is intelligible, i.e. if scientists understand the theory, they will have to have some idea of how the theory functions or how it produces certain outputs. Since de Regt allows for a wide variety of theories (for various degrees of exactness), he allows for a variety of criteria to assess the intelligibility of a theory.[11] He offers one possible criterion:

> **$CIT_1$:** A scientific theory *T* (in one or more of its representations) is intelligible for scientists (in context *C*) if they can recognize qualitatively characteristic consequences of *T* without performing exact calculations.[12]

By including the individual scientists and the specific context, $CIT_1$ accommodates the pragmatic and context-dependent nature of the intelligibility of theories, and, hence, also of UT and UP, since both notions depend on the intelligibility of theories. Besides the particular qualities of the theory in question, the combination of established scientific practices in a certain field, the developed abilities or skills of the individual scientists, and the established and available background knowledge determine whether a theory is intelligible for an individual scientist or group of scientists, or not. Certain qualities of a theory provide tools only if the skills of a scientist are attuned to these qualities.[13]

The two criteria proposed by de Regt, CUP and $CIT_1$, constitute the basis of his philosophical theory of understanding. Understanding, explanation, and prediction are viewed to be interrelated goals of science. He claims that prediction is not possible without understanding. If a scientist does not understand the important features and structure of a theory, if the theory is not intelligible to her, she will not be able to produce successful predictions. Instead, she will merely be guessing. A successful prediction that can be explained by a scientist shows that she understood a

9 See Humphreys (2000), p. 269.

10 See de Regt (2017), pp. 99f.

11 See ibid. pp. 101f.

12 Ibid. p. 102.

13 See ibid. p. 103.

theory. Hence, predictions enter de Regt's account of understanding through $CIT_1$, according to which scientists must be able to recognize qualitative characteristic consequences of a theory. And crucially, both kinds of understanding, UT as well UP, cannot be achieved by performing a rule-following procedure. Instead, tacit skills, the know-how to make use of a theory or an explanation, are required. Which skills a scientist needs to make a theory intelligible to her depends on the qualities of the theory. By applying $CIT_1$, it is possible to check whether the appropriate skills for a specific theory are present.[14]

In making decisions or performing reasoning, humans often refer to intuitions. Intuitions, in de Regt's view, can be defined as judgements whose underlying reasons are not fully accessible for a subject and which are results of heuristics that have been developed in an evolutionary process. The human cognitive system has produced these heuristics in response to environmental influences. It is possible to develop reliable intuitive skills in an environment that is sufficiently stable to make successful predictions if a subject has the opportunity to learn these regularities through practice. These conditions are usually fulfilled in science. Therefore, the skills acquired by scientists allow them to intuitively recognize theoretical consequences of a theory. For achieving scientific understanding, skills and intuitive judgements are crucial.[15]

> If a theory is intelligible to scientists because its theoretical qualities match their skills, they can reason "intuitively" with it. Like our everyday intuitive skills, scientists' skills are the outcome of a complex learning process in which they find themselves (that is, the historical and disciplinary context of their science).[16]

As de Regt has already mentioned, the construction of models or of explanations more generally is a matter of skill, of pragmatic decisions which lead to the desired result. Scientists have to have the know-how, the ability, to address and solve a new problem.[17] There are no fixed general rules that guide every possible construction process, or so many different rules that it is impossible to pick the right ones out of a big catalogue. Various theories as well as models of scientific explanation provide different tools for understanding, and all of them might be legitimately used in certain circumstances or contexts.[18] Hence, de Regt is an explanatory pluralist.

De Regt wants to clarify in the framework of his theory how precisely understanding and intelligibility are related to certain contexts. In short, the context determines which tools are available or deemed suitable. Which contextual factors are

14 See ibid. p. 107f.
15 See ibid. pp. 109ff.
16 Ibid. p 110.
17 See ibid. p. 112.
18 See ibid. p. 115.

important? According to de Regt's theory, certain qualities of a theory are conducive to its empirical success because they enhance the intelligibility of a theory for those scientists who possess the skills to use the conceptual tools that are associated with these qualities. Since science is a community activity and scientific understanding is a community achievement, the conceptual tools are mostly established at the level of scientific communities (meso-level). In a sense, understanding a theory means to become familiar with it. As soon as a scientist has developed the relevant skills to use the theory in an intuitive way (CIT), the theory is familiar to her. More specifically, the tools have to be familiar to the users. If scientists have developed the relevant intuitive skills to use these tools, the tools are familiar to the scientists and they are able to make successful predictions, which improves the intelligibility of the theory.[19]

Within the literature on scientific explanation, one finds pragmatic theories of explanation, which also allow for a plurality of explanatory strategies in scientific practice.[20] Such theories are based on the idea "that explanations are given and received by particular people in particular contexts for particular purposes. Different contexts, people, and purposes may require different types of explanation."[21] De Regt focuses on the pragmatic theory of Bas van Fraassen[22], who considers explanations to be answers to why-questions. But a why-question alone cannot determine what kind of answer is asked for. Additionally, the context has to be taken into consideration to make sense of the question and to know what kind of answer is adequate. Philip Kitcher and Wesley Salmon[23] have criticized van Fraassen's theory because he does not give criteria on when a factor is explanatorily relevant. Although van Fraassen states that explanatory relevance requires scientific relevance and explicates under which conditions an answer is scientifically relevant, he also states that not all scientifically relevant factors are explanatorily relevant, which leaves the problem unsolved. The context determines which scientifically relevant factors are also explanatorily relevant. By adding the two basic epistemic values, de Regt wants to solve van Fraassen's problem. Only if an explanation conforms to internal consistency and empirical adequacy, will it be explanatorily relevant.[24]

Summing up, de Regt covers two different kinds of scientific understanding with his theory, which are UT and UP. Scientists need to have intelligible theories, need to understand theories, if they want to understand phenomena. On the basis of an intelligible theory, scientists can construct explanations of phenomena, and

19 See ibid. pp. 116–119.

20 See ibid. p. 123.

21 Ibid. p. 124.

22 See van Fraassen, B. C. (1980), *The Scientific Image*. Oxford, Clarendon Press, DOI: 10.1093/0198244274.001.0001, especially chapter 5.

23 See Kitcher, P. & Salmon, W. C. (1987), "Van Fraassen on explanation." *Journal of Philosophy*, 84 (6), pp. 315–330, DOI: 10.2307/2026782.

24 See de Regt (2017), pp. 125–128.

hence understand them. Whether any theory is intelligible for some scientist and which type of explanation is constructed depends on the context, the skills of the individual researcher as well as the practices accepted by the respective community. De Regt presents three extensive case studies from the history of physics to substantiate and illustrate his theory of understanding.

The analysis of scientific understanding provided by Henk de Regt focusses on explanatory understanding, which is legitimate, but he does not give explicit arguments on whether, and if so why, explanatory understanding is the most important or general kind of understanding or how explanatory understanding is related or incorporated into other forms of understanding. In contrast to de Regt, Kareem Khalifa explicitly addresses the relation between explanatory and objectual understanding. So, let us have a look at Khalifa's account of scientific understanding, which varies significantly from de Regt's theory in several ways.

## 2.2 Kareem Khalifa: Scientific understanding is scientific knowledge of an explanation

Khalifa develops a model of understanding that he labels EKS model of understanding (explanation, knowledge, science model), since these three concepts are crucial for his account. Before talking about the details of this model of understanding, it is important to recognize two crucial features of it: First, Khalifa explicitly includes the fact that understanding is gradual in his account. That is, he provides a framework that incorporates the fact that understanding comes in degrees. De Regt's theory does not accommodate this. Khalifa's theory of understanding shall allow for the possibility to compare the understanding of different individuals, he develops a comparative account of understanding, while he does not provide a quantitative analysis of degrees of understanding. This is probably impossible. Rather, he takes it to be sufficient that in certain situations it is possible to decide which subject has a better understanding.[25] Second, Khalifa is only concerned with explanatory understanding, or understanding-why.[26] This is a common aspect of the theories of de Regt and Khalifa.

His principle of better understanding, how Khalifa calls it, takes the following form:

25 See Khalifa (2017b), chapter 1.

26 See ibid. pp.2f.

*EKS₁*: $S_1$ understands *why p* better than $S_2$ if and only if:

(A) *Ceteris paribus*, $S_1$ grasps *p*'s explanatory nexus more completely than $S_2$; or

(B) *Ceteris paribus*, $S_1$'s grasp of *p*'s explanatory nexus bears greater resemblance to scientific knowledge than $S_2$'s.[27]

Let's consider the EKS-model in more detail. The first principle (A) is called *Nexus Principle*. Khalifa starts with the idea that the subject's understanding of a phenomenon increases if she knows more correct explanatory factors that contribute to the phenomenon, and if she knows more of the relations that exist between these factors. On this basis, Khalifa defines the explanatory nexus of a phenomenon *p* as "the set of correct explanations of *p* as well as the relations between those explanations."[28]

If the explanatory nexus of *p* only includes correct explanations, how can it be determined whether an explanation is correct? Khalifa presents these four conditions:

*q* (correctly) explains *p* if and only if:

(1) *p* is (approximately) true

(2) *q* makes a difference to *p*

(3) *q* satisfies your ontological commitments (so long as they are reasonable); and

(4) *q* satisfies the appropriate local constraints.[29]

Notice that *q* denotes the explanans, the statement that does the explaining of *p*. The fourth condition is crucial: like de Regt, Khalifa explicitly allows for an explanatory pluralism. He does not give a strict definition of explanation. In fact, he even allows to identify 'explanation' with 'explanatory information'. With local constraints he refers to the specific interest of the researcher, the established standards of the discipline, and so on. Local constraints are context-dependent. Like de Regt, Khalifa wants to formulate an account of understanding that is universally valid, but allows for contextual variation. Khalifa reaches this goal by formulating three global conditions and one local condition for understanding.[30]

The explanations belonging to one explanatory nexus can stand in in many varying relations to each other, and the grasp of an explanatory nexus is more complete if more explanations and inter-explanatory relations are grasped, if the quality or importance of the explanations and inter-explanatory relations are grasped, or if more details of the explanations and inter-explanatory relations are grasped. Again,

27 Ibid. p. 14.

28 Ibid. p. 6.

29 Ibid. p. 7.

30 See ibid. pp. 6ff.

this model of understanding is not supposed to offer a quantitative analysis of understanding. All the dimensions just mentioned, quantity, quality and the level of detail of explanations and inter-explanatory relations, figure into understanding, and it depends on the specific context or situation which dimension of one's grasp is more important.[31]

The second principle (B) is the *scientific knowledge principle*. This principle captures everything Khalifa takes to be necessary for a characterization of grasping. He defines grasping as "a cognitive state bearing some resemblance to scientific knowledge of some part of the explanatory nexus."[32] This implies the question: what is scientific knowledge? Khalifa offers this definition: "*S* has scientific knowledge that *q* explains *why p* if and only if the safety of *S*'s belief that *q* explains *why p* is because of her scientific explanatory evaluation."[33] In short, knowledge counts as scientific knowledge if it has been gained by a scientific explanatory evaluation, SEEing. According to Khalifa, SEEing consists of three components: the consideration of plausible potential explanations of the phenomenon of interest, the comparison of the potential explanations, and finally of the formation of (doxastic) attitudes based on the comparisons. SEEing ensures the safety of one's explanatory commitments and therefore the status of this kind of knowledge as scientific. The grasp of a subject bears greater resemblance to scientific knowledge along the following dimensions: the number of plausible potential explanations the agent has considered, the number of considered explanations that have been compared using scientifically acceptable methods and evidence, the scientific status of these methods and evidence that has been used, the safety of the agent's beliefs about explanations, the accuracy of the agents beliefs about explanations, and finally the variety of ways that the agent can use explanatory information so as to achieve different scientific goals.[34]

To avoid that principles (A) and (B) of the EKS-model could come into conflict with one another, the *ceteris paribus* clause is used.[35] Either $S_1$ grasps more items of *p*'s explanatory nexus (i.e. she knows more items that belong to the nexus) than $S_2$, while both are equally competent in a specific scientific field, or the grasp (i.e. knowledge) of $S_1$ is more scientific than the grasp of $S_2$, e.g. if $S_1$ is a professor in a certain domain and $S_2$ a very interested lay person.

In sum, Khalifa's account of scientific understanding stands in the tradition of the "received view" of understanding, as Khalifa calls it,[36] and he provides this definition of the received view:

31 See ibid. pp. 9f.

32 Ibid. p. 11.

33 Ibid. p. 12.

34 See ibid. p. 11ff.

35 See ibid. p. 15.

36 See ibid. p. 16.

> *S* understands *why p* if and only if there exists some *q* such that *S* knows that *q* explains *why p*.[37]

That is Khalifa's view on scientific understanding. Let us now look at a third alternative.

## 2.3 Finnur Dellsén: Understanding as dependency modelling

The third account I want to introduce is the one offered by Finnur Dellsén. His account differs significantly from the other two accounts just presented, since Dellsén argues for understanding without explanation, namely an account of objectual understanding. To begin with, Dellsén points to some features of objectual understanding. First, he specifies objectual understanding as understanding of phenomena. Dellsén is not interested in understanding topics, subject matters or theories, as some other advocates of objectual understanding do[38], because he wants to avoid the slip form understanding something to understanding a discipline that studies or an account that refers to the thing or phenomenon that is to be understood. Second, understanding is gradual in a way that propositional knowledge is not. Subjects can understand a phenomenon to different degrees and the degree of understanding of a phenomenon of one subject can change over time. This is a widely, I would even say universally, accepted feature of understanding. And third, Dellsén assumes that paradigmatic cases of objectual understanding can be found in the empirical sciences. Therefore, he takes his account to be an account of scientific understanding.[39] Given these characterizations of the kind of understanding that Dellsén is concerned with, it is justified to assume that Dellsén is dealing with the same kind of understanding as de Regt and Khalifa: the understanding of phenomena achieved by scientists.

Dellsén's "account of understanding [...] holds that to understand a phenomenon is to grasp a specific kind of model of that phenomenon's dependence relations."[40] He calls his account 'dependence modelling account', DMA for short. Models, in

---

37 Ibid. p. 18.

38 Like Christoph Baumberger for instance, see Baumberger, C. (2011), "Types of Understanding: Their Nature and Their Relation to Knowledge." *Conceptus*, 40, pp. 67–88, DOI: 10.1515/cpt-2014-0002; and Baumberger, C. & Brun, G. (2017), "Dimensions of Objectual Understanding." In Grimm S. R., Baumberger, C. & Ammon, S. (eds.), *Explaining Understanding. New Perspectives from Epistemology and Philosophy of Science*, pp. 165–189, New York and London, Routledge.

39 See Dellsén, F. (2020), "Beyond Explanation: Understanding as Dependency Modelling." *Brit. J. Phil. Sci.*, 71, pp. 1261–1286, DOI: 10.1093/bjps/axy058, pp. 1263f.

40 Ibid. pp. 1264f.

Dellsén's view, are information structures of some kind that are interpreted so as to represent the target.

> These information structures can be concrete, as in Watson and Crick's original model of DNA, or abstract, as in mathematical or computational models like the Lotka-Volterra model of predation in ecological systems. In both cases, the structures are associated with an intended interpretation that specifies how the different parts of the structure correspond to different elements and relations in the phenomenon – a 'key'.[41]

Importantly, no information structure, may it be a material object or a system of equations, is a model in itself. Only through the interpretation of (parts of) these concrete or abstract information structures do they become part of a model.[42] Parts of the information structure must be associated with specific parts of the phenomenon in order to be a model of that phenomenon. Additionally, as understanding takes place in the mind of individuals, the models that are used for understanding a phenomenon must somehow be related to (human) thought. Dellsén does not analyze in detail what the relation between models and mind might be, but rather uses the term 'grasp' for referring to this relation.[43] That is, in the opinion of Dellsén, "understanding consists of grasping a certain kind of model of the understood phenomenon."[44]

"What kind of model must an understanding agent grasp?"[45] Since models are always incomplete or inaccurate representations, which aspects of a phenomenon must be represented in a model that enable understanding? Based on former work that relates understanding-why with dependence relations, Dellsén states "that the aspects of a phenomenon that matter for understanding are the dependence relations that the phenomenon, or its features, stands in towards other things."[46] He

---

41 Ibid. p. 1265. Dellsén uses the notion of a 'key' from Frigg and Nguyen, see Frigg, R. & Nguyen, J. (2016), "The Fiction View of Models Reloaded." *The Monist*, 99, pp. 225–242, DOI: 10.1093/monist/onw002.

42 Dellsén is following Weisberg in viewing models as interpreted structures. The kind of modelling that Dellsén is thinking of resembles Weisberg's 'target-directed modelling'. For more information, see Weisberg, M. (2013), *Simulation and Similarity. Using models to understand the world*, Oxford, Oxford University Press DOI: 10.1093/acprof:oso/9780199933662.001.0001, especially chapter 5.

43 See ibid. p. 1265.

44 Ibid. p. 1265.

45 Ibid. p. 1266.

46 Ibid. p. 1266. Dellsén is referring here to Kim, J. (1974), "Noncausal Connections." *Noûs*, 8, pp. 41–52, DOI: 10.2307/2214644; and Greco, J. (2014), "Episteme: Knowledge and Understanding." In Timpe, K. & Boyd, C. A. (eds.), *Virtues and Their Vices*, pp. 285–302, Oxford, Oxford University Press, DOI: 10.1093/acprof:oso/9780199645541.003.0014; and Grimm, S. (2014), "Un-

does not limit the notion of dependence relations to causality, but allows for various kinds of dependence relations to hold between (parts of) phenomena, for example grounding relations or mathematical relations.[47] "The kind of model that [he] think[s] is involved in understanding is one that aims to capture the network of dependence relations that a phenomenon stands in, whatever these relations turn out to be. [He] will refer to this as a 'dependency model'."[48]

However, Dellsén does not take grasping any dependency model of a phenomenon as being enough for understanding that phenomenon. "Rather, the model must in some sense be a 'good' representation of the relevant dependence relations."[49] The straightforward suggestion is that the quality of a model depends on the extent to which the model correctly depicts the network of dependence relations of a phenomenon. Yet, a dependency model can fail in two respects in depicting the network of dependence relations, either by incorrectly representing (misrepresenting) some aspects of the network (idealization) or by not representing certain aspects at all (abstraction). Hence, Dellsén recognizes two distinct criteria: accuracy and comprehensiveness. These criteria might come into conflict, the increase of the one might require the sacrifice of the other. As understanding depends on both criteria, it is possible to increase one's understanding by sacrificing one of the two criteria sometimes, given that this brings sufficient benefit in terms of the other criterion, according to Dellsén. Thus, his model-based account can explain the gradual nature of understanding in terms of two other gradable notions, namely accuracy and comprehensiveness.[50]

A final important concept in Dellsén's account of understanding is context, which has several functions. First, any context determines a threshold for the degree of understanding. Attributing understanding of a phenomenon to a subject requires that the subject grasps a sufficiently accurate and comprehensive model of the phenomenon, so that the understanding exceeds the contextually determined threshold. Second, any context specifies which parts of a complex phenomenon have to be sufficiently understood in order to understand the phenomenon itself. Ecologists, physicians and psychologists will all understand human mating, but in different ways. And third, the context also designates which (parts of) other phenomena are striking enough so that their dependence relations to the target

derstanding as Knowledge of Causes." In Fairweather, A. (ed.), *Virtue Epistemology Naturalized: Bridges between Virtue Epistemology and Philosophy of Science*, pp. 347–360, Dordrecht, Springer, DOI: 10.1007/978-3-319-04672-3_19. Dellsén argues that objectual understanding can come apart from explanation, which is not possible in the view of Greco and Grimm, at least not according to Dellsén's reading of their work.

47 See ibid. p. 1266.

48 Ibid. p. 1266.

49 Ibid. p. 1267.

50 See ibid. p. 1267.

phenomenon, or lack thereof, become relevant for the understanding of the target phenomenon. For example, one does not need to consider the whole history of western societies to understand the length of the shadow of a flagpole at a certain time, but one should have some basic knowledge from physics and geometry.[51]

In short, Dellsén proposes the following dependency modelling account (DMA) of understanding:

> DMA: *S* understands a phenomenon, *P*, if and only if *S* grasps a sufficiently accurate and comprehensive dependency model of *P* (or its contextually relevant parts); *S*'s degree of understanding of *P* is proportional to the accuracy and comprehensiveness of that dependency model of *P* (or its contextually relevant parts).[52]

DMA does not require explanation, although Dellsén takes dependence relations to usually undergird explanations. He contrasts his DMA with explanatory accounts of understanding, which he summarizes in the following way:

> *U→E*: *S* understands *P* only if *S* grasps enough of an adequate explanation of *P* (or its relevant features); other things being equal, *S* has more understanding of *P* to the extent that *S* grasps more of an adequate explanation of *P* (or its relevant features).[53]

*U→E* is intended to capture any account of explanatory understanding that takes explanation as a necessary requirement for understanding. Dellsén then discusses three cases in which, according to him, *U→E* fails to accommodate the understanding that scientists achieve, whereas DMA can cope with such types of cases. I will engage with these three cases in detail in section 6.2.3, but now, let's recap the three different accounts of scientific understanding.

## 2.4 Two questions concerning scientific understanding

So, what shall we make of these three different accounts of scientific understanding? All of them share some common ground. First, they are all intended to conceptualize scientific understanding as an understanding that is gained in *science in general*, or in science as a whole, and do not distinguish between different scientific disciplines. Some fundamental unity of science seems to be assumed by de Regt, Khalifa, and Dellsén. In this book, I take up this assumption and will be concerned with a general

---

51 See ibid. pp. 1267f.

52 Ibid. p. 1268.

53 Ibid. p. 1269.

account of scientific understanding that will be able to accommodate all scientific disciplines, or as many as possible, by pointing out commonalities, while still leaving room for differences between disciplines. Second, however, the criteria, principles, or definitions of understanding presented by the three scholars refer to understanding that *individual* scientists achieve. The only exception is de Regt's CUP, which does not refer to individuals specifically. However, if one takes a look at the case studies that de Regt provides, one will see that he is investigating whether certain individual scientists understood some phenomenon, not only some theory. While de Regt does mention that science can be divided into a macro-, meso-, and micro-level, he does not analyze in which regards understanding of phenomena differs with respects to these three levels. So, all three scholars introduced in this chapter take understanding to be some achievement that can be gained or realized (primarily) by individuals. Third, all accounts address understanding of phenomena achieved in science or take understanding of phenomena as an ultimate aim of science. While other types of understanding might be necessary for understanding phenomena, like the understanding of theories or models used in research, the most important and interesting type of understanding in science is the understanding of the phenomena that are investigated. Hence, I will focus on the understanding of phenomena, too. And fourth, all three scholars agree that understanding is context-dependent, that the historical and disciplinary circumstances or local constraints have an impact on the achievement of understanding and the assessment of its quality. I will take this insight into account as well.

So much for the agreement and commonalities of the three accounts. Yet, they also differ in crucial respects. For instance, de Regt takes scientific understanding to be some kind of ability or know-how, since he argues that scientists need to *be able* to use a theory to construct an explanation of a phenomenon. If scientists are unable to construct an explanation, they will not have understood the phenomenon. Khalifa, in contrast, claims that scientific understanding is scientific knowledge of an explanation, and not an ability. And in Dellsén's view, scientific understanding is the ability to grasp a 'good' dependency model of some phenomenon, an ability that does not require explanation. So, de Regt and Khalifa agree that understanding requires explanation, a feature that is denied by Dellsén. However, de Regt and Dellsén both take understanding to be an ability and not a kind of knowledge, as Khalifa explicitly claims. The comparison of these accounts highlights that at least two questions about scientific understanding are of central interest, but not ultimately resolved within the debate. These two questions are:

1) Does scientific understanding require explanation or not?
2) Is understanding an ability or a type of knowledge?

In order to answer the main research questions of this book, what scientific understanding is and how scientists achieve it, I have to address the two questions identified through the comparison of the three accounts of understanding. That is because answers to my main research questions will depend on the answers one gives to the other two questions just mentioned. If understanding requires explanation, some explanation must somehow be involved or related to the understanding, may it be as a starting point or a product. If understanding does not require explanation, scientists will not have to draw on or produce any explanation for understanding, and hence might acquire the latter in quite a different way than if some explanation would be involved. Furthermore, if understanding is some ability, a type of knowing-how, it might have quite different characteristics than if it would be some form of propositional knowledge or knowledge-that. Moreover, the acquisition processes of knowing-how and knowing-that are very different from one another, as various scholars have already pointed out. This issue will be the topic of section 4.1.

So, in sum, depending on whether scientific understanding requires explanation in some way or not and whether understanding is an ability or a form of propositional knowledge, the way how scientists actually acquire understanding of phenomena will turn out to be fairly different. Hence, it is necessary to first provide answers to the questions concerning the relation of understanding and explanation and the nature of understanding, that is, whether understanding is an ability or a type of knowledge. Only then can the third question, how scientists actually achieve scientific understanding, be answered. Thus, these three questions are the ones I am going to answer in this dissertation. I will start with the question concerning the necessity of explanation for understanding, which will be answered in chapter three. I then turn to the nature of understanding in chapter four. The third question, which I take to be the most interesting one, will be answered in the course of chapters five and six. But first things first, let us start with looking at explanation and its potential role in understanding.

# 3. Scientific understanding, scientific explanation, and why they cannot be torn apart

The connection between understanding and explanation is a core topic in the debate about understanding, especially for philosophers of science. The presentation of the accounts of scientific understanding developed by Henk de Regt, Kareem Khalifa and Finnur Dellsén in the previous chapter exemplify this claim. On the one hand, the accounts of understanding from de Regt as well as Khalifa are only concerned with explanatory understanding, a type of understanding that is tied to explanation. De Regt admits that kinds of understanding without explanation exist, but his account of understanding is not intended to cover these kinds. In contrast to de Regt, Khalifa explicitly argues that at least one other kind of understanding, namely objectual understanding, can be reduced to explanatory understanding. On the other hand, Dellsén presents an account of scientific understanding in terms of objectual understanding, according to which explanation is not necessary for understanding. What should we make of these different positions? Does scientific understanding require explanation or not?

In this chapter, I argue that scientific understanding does require explanation. To do so, I first address the concept of explanation. Explanation is one of the core concepts in philosophy of science and various accounts and definitions of explanation have been provided in the last decades. To avoid confusion, I need to clarify what I mean by the concept and the term 'explanation'. Hence, I start with a very brief discussion of explanation in section 3.1, in which I introduce the generic conception of explanation that I adopt throughout this book. I then turn to the main topic of this chapter, the relation of scientific understanding and explanation. I do so by discussing arguments that are proposed to support a view of understanding that is independent of explanation. In section 3.2, I present and critically discuss Peter Lipton's view on understanding without explanation. I analyze the four examples that Lipton provides as instances of understanding without explanation and argue that none of the examples succeeds in being an instance of understanding without explanation. Subsequently, I delve into the discussion about two forms of understanding that some authors strictly distinguish, namely, objectual and explanatory understanding. In section 3.3, I present the view of Jonathan Kvanvig, who argues that explanatory and objectual understanding are intrinsically different and the counterarguments from Khalifa against Kvanvig. I argue that Khalifa's critique of Kvanvig's conception of objectual understanding is in line with my criticism on Lipton's view. As the accounts from Lipton and Kvanvig are unrelated while facing the same problems, a conception of scientific understanding without explanation becomes more and more implausible. Finally, in section 3.4, I engage with further arguments in favor of and against a separation of objectual and explanatory understanding. Christoph Baumberger wants to distinguish objectual and explanatory understanding in terms of their targets and vehicles. Following Stephen Grimm's argument why a distinction in terms of the targets of understanding is not possible, I will argue that the distinction in terms of the vehicle is not possible either. I conclude that, at least for scientific understanding of phenomena, a differentiation between objectual and explanatory understanding is not reasonable, as both terms, in their prevalent use, cannot accommodate scientific practice and the function of explanation within it. Hence, scientific understanding is not possible without explanation.

One important remark is necessary before the analysis of the relation of understanding and explanation. Although I am exclusively dealing with scientific understanding of phenomena, understanding that is gained in the scientific domain, many authors in the debate are concerned with understanding in general and do not reduce their analysis to scientific understanding. The controversies about understanding and explanation, about objectual and explanatory understanding, which I examine in this chapter, are also not restricted to scientific understanding. However, this is not a problem for my project. I analyze the plausibility of arguments in favor of an independence of understanding from explanation for the scientific domain, whether it makes sense for science to conceptualize scientific understanding as being independent from explanation. I am not claiming that any type or kind of understanding requires explanation. In fact, I do think that there are types of understanding that are independent of explanation. But these types will not be typical or distinctive as an aim of science, so I shall argue. Therefore, any argument concerning the relation of understanding and explanation needs to be interpreted in light of scientific practice if scientific understanding as an aim of science is the target of investigation. Having this clarification in mind, I do not take it to be problematic that scientific understanding is not always clearly distinguished from other types of understanding by all authors.

## 3.1 A few words on explanation

Prior to delving into the discussion of whether understanding requires explanation, some considerations concerning the concept 'explanation' are necessary. Explanation has been and still is one of the most central concepts in philosophy of science.

As Henk de Regt nicely puts it, even "after sixty years of debate about scientific explanation, there is currently no consensus favoring one model but rather a plethora of different models of scientific explanation."[1] Among the types of explanation that are proposed and discussed are deductive-nomological, unificationist, model-based, causal, counterfactual, mechanistic, functional, probabilistic, or mathematical explanations, and this list is not exhaustive. Some of these types overlap, some can or might be reduced to another type.[2] Since I am concerned with scientific understanding, and not with scientific explanation, I do not attempt to develop and provide a specific conception of scientific explanation. This issue would be more than enough for another research project. However, I do adopt a generic conception of explanation. In this section, I delineate the basic features of this generic conception of explanation.

What is an explanation? Attempts to answer this question led to the emergence of two main opposing camps: adherents of an ontic conception of explanation versus proponents of an epistemic conception. While according to the ontic conception explanations are things or facts that exist or take place in the world, the epistemic conception suggests that explanations are (complexes of) representations of things or facts in the world. Consequently, for the ontic conception explanations exist independently of any cognitive subjects, whereas the epistemic conception requires subjects to construct explanations, representations, of things in the world. As no decisive argument in favor of or against one of the two conceptions could be provided so far, I follow my intuition and adopt an epistemic conception of explanation. In my view, it is more plausible to speak of things like entities, phenomena, events, or structures to be in the world, while explanations are constructed to represent these things. Explanations are created by subjects and if there were no subjects trying to explain things in the world, these things would still exist, but there would be no explanations.[3]

So, for the purpose of this book, explanations are representations. The next question is what makes a representation an explanation and not merely a description. In this regard, I follow Hayne Reese and, very roughly, view explanations to pro-

1 De Regt (2017), p. 49.

2 For an overview on the different types of scientific explanation, see for example Woodward, J. & Ross, L., "Scientific Explanation", *The Stanford Encyclopedia of Philosophy* (Summer 2021 Edition), Edward N. Zalta (ed.), URL = https://plato.stanford.edu/archives/sum2021/entries/scientific-explanation/ (last accessed April 12th, 2022).

3 For one line of argumentation against the ontic and in favor of the epistemic conception of explanation, see Wright, C. & van Eck, D. (2018), "Ontic Explanation Is either Ontic or Explanatory, but Not Both." *Ergo: An Open Access Journal of Philosophy*, 5, pp. 997–1029, DOI: 10.3998/ergo.12405314.0005.038.

vide reasons for *why* something is the case or could be the case,[4] while descriptions merely state what *is* the case. Descriptions provide us with facts (e.g. the sky is blue and blue light is scattered more than other colors by the atmosphere), while explanations give reasons for these facts (the sky is blue, *because* blue light is scattered more than other colors by the atmosphere). An explanation transcends a description, as a representation "becomes explanatory because it goes beyond the question answered by description – "What happens?" – to the question answered by explanation – "Why does it happen?"".[5] The notion of reasons is deliberately kept vague, in order to capture all kinds of reasons that are deemed crucial or adequate in different contexts. Reasons include causes, but they are not limited to causes, as not all types of explanation refer to an actual cause of a phenomenon. Famously, unificationist explanations provide unified accounts of various different phenomena by deriving descriptions of as many different phenomena as possible from as few argument patterns as possible, to use Philip Kitcher's technical vocabulary, but without referring to any actual causes of some phenomenon.[6]

De Regt and Khalifa, despite their differences with respect to understanding, agree on one crucial aspect: they allow for an explanatory pluralism to achieve understanding. Both authors argue that, depending on the historical or disciplinary context, various explanatory strategies lead to understanding. Following a review of various types of explanation, de Regt "conclude[s] that understanding is a universal aim of science that can be achieved by contextually varying modes of explanation."[7] And Khalifa claims that an explanation must "satisfy "local constraints" [...] [as] the relevance of many explanatory features depends on the specific explanandum, the standards of the discipline, and the interest of the inquirer."[8] Hence, there is not one kind of explanation that provides the best or most accurate understanding in

---

4 I am referring here to the concept of how-possibly explanations, explanations that do not explain why something actually happened, but rather explain how something is or was possible. Yet, whether how-possibly explanations should be seen or treated as genuine explanations is a contested question, which I will not address here. For more information, see for example Brainard, L. (2020), "How to Explain How-Possibly." *Philosophers Imprint*, 20 (13), pp. 1–23; or Reydon, T. (2012), "How-possibly explanations as genuine explanations and helpful heuristics: A comment on Forber." *Studies in History and Philosophy of Biological and Biomedical Sciences*, 43 (1), pp. 302–310, DOI: 10.1016/j.shpsc.2011.10.015.

5 Reese, H. W. (1999), "Explanation Is Not Description." *Behavioral Development Bulletin*, 8 (1), pp. 3–7, DOI: 10.1037/h0100524, p. 4.

6 For Kitcher's account of unificationist explanation, see e.g. Kitcher, P. (1989), "Explanatory Unification and the Causal Structure of the World." In Kitcher, P. & Salmon, W. (eds.), *Scientific Explanation*, Minnesota Studies in the Philosophy of Science, Vol. 13, pp. 410–505, Minneapolis (MN), University of Minnesota Press.

7 De Regt (2017), p. 86. His full review of different models of explanation can be found ibid. chapter 3.

8 Khalifa, (2017b), p. 8.

all cases. Instead, the subjects involved in a process of gaining understanding must assess, according to the relevant standards that they follow, which explanation is the 'best' or appropriate one to lead to understanding in the respective context. Since the pluralist positions concerning explanation of de Regt and Khalifa are supported by a growing attention and literature on explanatory pluralism and diversity in science,[9] I adopt a pluralist position of scientific explanation as well. While being an explanatory pluralist, I do think that a generic conception of explanation, which leaves room for the various types of scientific explanation that can be found in scientific practice, can be articulated.[10] Hence, I provide the following conception of explanation:

> An explanation is a representation of relations of (parts of) the phenomenon under investigation, which provides reasons (an explanans) for features of (parts of) the phenomenon (the explanandum).[11]

This is the generic conception of explanation that I use and refer to when I speak about explanation in the remainder of this book. Again, I use the notion 'reasons' in this conception in order to include non-causal explanations, like unificationist, law-based, probabilistic, logical and further types of explanations. Having clarified what I mean by explanation, we can now turn to the actual topic of this chapter. Namely, the relation of understanding and explanation.

## 3.2 Cases of understanding without explanation?

Some philosophers in the debate on understanding maintain that in some cases, understanding can be gained without explanation. Peter Lipton is one of them. I

---

9 See for example Mantzavinos, C. (2016), *Explanatory Pluralism*. Cambridge, Cambridge University Press, DOI: 10.1017/CBO9781316440599; or Braillard, P.-A. & Malaterre, C. (2015), *Explanation in Biology. An Enquiry into the Diversity of Explanatory Patterns in the Life Sciences.* In History, Philosophy and Theory of the Life Sciences, Dordrecht, Springer, DOI: 10.1007/978-94-017-9822-8; or Weber, E., de Regt, H. W. & van Eck, D. (2021), "Investigating the Unity and Disunity of Scientific Explanation." *Found Sci*, 26, pp. 1021–2024, DOI: 10.1007/s10699-020-09704-x; or Rice, C. & Rohwer, Y. (2021), "How to Reconcile a Unified Account of Explanation with Explanatory Diversity." *Found Sci*, 26, pp. 1025–1047, DOI: 10.1007/s10699-019-09647-y.

10 I got the idea of formulating and using a generic conception of explanation from de Regt, who also provides a generic conception of explanation, though a different one. Cf. de Regt (2017), pp. 24f.

11 Note that I am concerned with scientific understanding of phenomena that are the targets of scientific investigations in this book and that I will not analyze what it means to understand a theory scientifically. Hence, I stay agnostic as to whether this generic conception of explanation is applicable to understanding other objects than phenomena, like for example theories.

present Lipton's examples, which he simply calls causation, necessity, possibility, and unification, and argue that he fails to show that scientific understanding is possible without explanation. Either he is wrong in claiming that no explanation is involved in the discussed cases, or he does not make a convincing point that his arguments are applicable to or can accommodate scientific understanding.

Before I address Lipton's arguments that there are cases of understanding without explanation, I want to mention some general aspects concerning Lipton's view of understanding. First, he is not exclusively concerned with scientific understanding, or at least he does not say so explicitly. Hence, I take it that Lipton is engaged in the analysis of understanding more generally and I have to analyze whether his view is plausible for science. Second, Lipton identifies understanding with the cognitive benefits that an explanation provides. These cognitive benefits are, in turn, identified as kinds of knowledge, including knowledge of causes, of necessity, of possibility, and of unification. In short, Lipton takes understanding to be certain kinds of knowledge that are provided by explanations. Importantly, understanding is not identified with the explanation itself, but rather with its benefits. This point is crucial, as it enables a separation of understanding and explanation.[12] As I have not addressed the question of whether understanding should be conceptualized as a kind of knowledge (-that) or rather as an ability (knowledge-how), which I do in chapter four, I adopt Lipton's conception of understanding as being knowledge of causes etc. for the discussion of his cases and argue that it is implausible how subjects should gain the understanding Lipton attributes to them without explanation.

### 3.2.1 Causation

The first example presented by Lipton is causation. He identifies causal information as a form of understanding. Many explanations provide this kind of understanding, but Lipton wants to investigate whether it is possible to gain causal information without an explanation involved.

> We need cases that, in addition to not being explanations themselves, do not work by means of generating explanations that are then the proximate cause of the consequent understanding. [...] [In such cases,] the process of acquiring understanding does not begin with an explanation, but the understanding is nevertheless a product of an explanation, which is not what we are looking for here.[13]

---

12 See Lipton, P. (2009), "Understanding without explanation." In de Regt, H. W., Leonelli, S. & Eigner, K. (eds.), *Scientific Understanding: Philosophical Perspectives*, pp. 43–63, Pittsburgh, University of Pittsburgh Press, pp.43f.

13 Ibid. pp. 44f.

Lipton works with the premise that there is something like tacit understanding, but nothing like a tacit explanation. This assumption enables him to identify cases where causal information, alias understanding, can be achieved without the influence of any explanation. This happens via the use of images and physical models. Persons may grasp causal information that are provided by these devices, they may achieve genuine understanding, without being able to express an explanation that contains this information. Manipulation, in Lipton's view, is an even stronger example of understanding without explanation. A scientist may be an expert in using a complicated machinery because he acquired the relevant causal information, but he may not be able to explain this information to others. In sum, Lipton wants to be able to differentiate between someone who simply knows that a phenomenon occurs and someone who has a deep understanding of the causes of the phenomenon, but might not be able to verbalize the causal information. He also mentions a possible critique to this idea, namely that a person may at least be able to say something about the causes of a phenomenon, even if some causal information remains tacit, i.e. cannot be made explicit. In such a case, Lipton maintains, the person would be able to provide an explanation, but this explanation would not exhaust the understanding of the person. Therefore, parts of the understanding of the person still do not require explanation.[14]

I fail to see how understanding can be attributed in these cases, and also how understanding conceptualized in this way should be valuable for science. I will first address the case of images and models, second the case of manipulation, and finally Lipton's general point about the natures of understanding as being tacit and explanation as being verbal or explicit.

Images and models are created to convey information, to make certain features salient that might otherwise be hidden in the real phenomenon. But this information does not automatically pass on to the person. Every representation requires interpretation by the subject. Just by looking at a representation and not interpreting *what* is represented and *how* it is represented, the image or model will not provide understanding of the represented phenomenon for the subject. The user makes the image or model intelligible to herself only by interpreting the representation, interpretation is a crucial part of representation.[15] Interpretation requires some kind of reasoning about the object that is interpreted and relations of the object must be recognized. A user makes sense of the representation, recognizes the (causal) information captured in the model, by interpreting the model or image. And if some

14 See ibid. pp. 44ff.

15 The importance of interpretation is stressed in several philosophical accounts of representation, see for example Frigg, R. & Nguyen, J., "Scientific Representation", *The Stanford Encyclopedia of Philosophy* (Spring 2020 Edition), Edward N. Zalta (ed.), URL = https://plato.stanford.edu/archives/spr2020/entries/scientific-representation/ (last accessed April 12th, 2022).

causal information is represented, this relation will have to be recognized by the user and (correctly) interpreted. She must (correctly) identify what is the cause and what is the caused event shown in a representation. If the user achieves this, she will have a causal explanation of the represented phenomenon in her mind. For example, if a person sees an orrery, she will only gain understanding of planetary motion if she, first, identifies the model as a representation of the solar system (and not of an atomic nucleus orbited by electrons, for instance), and second, identifies the relations between parts of the representation. While parts of the process of interpretation might be tacit, I do not see how interpretation of representations is possible at all without recourse to some explicit conceptions that the person possesses. I engage with the relation of propositional or explicit knowledge and tacit knowledge or knowing-how in detail in sections 4.2 and 4.3. Additionally, it is not necessary that different people are able to give explanations that capture the same, or all, information. Representations can be good or bad, adequate or inadequate for certain purposes in certain contexts. People with different background knowledge might interpret a representation differently, their understanding of the representation might vary, but all of them would have gained some understanding, some causal knowledge, nevertheless.

I agree with Lipton that images and models provide information. But in the case of understanding, this information is not merely tacitly or unconsciously adopted and stored by the user. The information of a representation recognized by the user is consciously interpreted by her. If the user is not able to generate an explanation from an image or model, she has no understanding. Accordingly, she has not gained understanding without explanation, but she has rather not understood anything through the image or model, since she could not interpret the representation in light of her background knowledge. Lipton mentions the case of someone who "never *properly* understood the why of retrograde motion until [she] saw it demonstrated visually in a planetarium."[16] But this example implies that the subject knew about retrograde motion *before* she saw the visual representation and already possessed information about the planets apparently moving into an opposite direction, maybe even an explanation of retrograde motion, but she did not really understand the phenomenon merely on that basis. This is not the same as understanding retrograde motion by seeing it visually in a planetarium *without* having any explicit knowledge about it. And if the understanding provided by the model of retrograde motion is tacit understanding that cannot be made explicit, as Lipton argues, how would the person or anyone else be able to judge or to know that her understanding improved or is proper after seeing the visual model? Understanding the cause or an aspect of a phenomenon *properly* implies that the cause or aspect of the phenomenon must

16 Lipton (2009), p. 45, my emphasis.

be understood in a certain, proper, manner through a representation. If the understanding cannot be made explicit at all, it will not be possible to determine whether the person in question acquired a proper understanding, an inappropriate understanding, or maybe even no understanding at all, since tacit understanding is inaccessible for any subject, including the subjects that gained this understanding. If a person wants to make sure that she gained some understanding (causal knowledge) about a phenomenon by looking at a visual representation, she will have to make explicit what she understood.

Furthermore, according to my generic conception of explanation, which states that an explanation is a representation of relations of (parts of) the phenomenon under investigation, which provides reasons (an explanans) for features of (parts of) the phenomenon (the explanandum), images and physical models, the representations Lipton mentions, can be viewed as being explanations *themselves*. I do not restrict my conception of explanation to propositions. The same information concerning aspects of phenomena and their reasons can be captured in form of a proposition, an image, or a physical model, at least in many cases. Lipton apparently does not include images or physical models into his conception of explanation. I grant Lipton that models or images, in case they are not considered to be explanations themselves, can enable genuine understanding that is not possible by merely knowing a propositional explanation. This is a good point for arguing in favor of a genuine difference between knowledge of an explanation and understanding, but he does not show that understanding merely through visual representation and *without an (explicit) explanation at all* is possible. The visual representation of retrograde motion *alone* will not have provided understanding for the subject, since she would not have been able to make any sense of the representation without already knowing what is represented, and hence being able to identify the explanandum, the explanans, and their relation in this case.[17]

17 Victor Gijsbers is also not convinced by Lipton's example of images and models and his criticism is quite similar to mine. "Evidently, simply *seeing that* the planets perform a retrograde motion is not enough to count as understanding—if it did, we would not even need the planetarium, but could just look at the night sky. What more is needed? Well, we should be able to identify the salient features of the system, the features that determine that retrograde motion occurs. [...] Anyone who hasn't grasped that the fact that the earth's shorter sidereal period is essential to the appearing of retrograde motion, has not understood why the phenomenon occurs. But anyone who has grasped this possesses an explanation of apparent retrograde motion. If the person were not able to express this explanation to others, perhaps because of a lack of useful vocabulary or linguistic skills in general, it would be pedantry to say that he understands but cannot explain. Even if he cannot express it, he does have an explanation." Gijsbers, V. (2013), "Understanding, explanation, and unification." *Studies in the History and Philosophy of Science*, 44, pp. 516–522, DOI: 10.1016/j.shpsa.2012.12.003, p. 518, original emphasis.

The case of manipulation is as problematic as the case of representation. How should it be possible to attribute genuine understanding of a machinery to a person who is not able to explain what the machinery does when she uses it? Simply using the machinery without the ability to explain at least parts of the processes is identical to the stump, automatic behavior of robots, who perform their tasks exactly by following rules without understanding what they do or why they are doing something precisely in that way and not another. When an agent really comes to understand a complicated machinery through using it, for example a complex laser system, it will be a trial and error process in the beginning. She will figure out what happens if she does certain things with the system. By continuing, she will be able to reason which actions produce which effects, she will understand it and articulate the relations. Again, it is not possible to reasonably attribute a sophisticated understanding, which is what Lipton wants to do, to someone who is able to manipulate a machinery very accurately and in every possible respect, but who is not able to articulate and explain in any way what is happening. Merely keeping a machine running does not amount to a genuine understanding of that machinery. Imagine the case of two scientists, Amy and Bob, who work with precisely the same laser system. Amy is able to explain that she gets a clear signal out of the system when all the mirrors are in a certain position, because this configuration ensures that all the light beams are in phase and, therefore, amplify the signal. Bob, in contrast, can only say that he gets a clear signal out of the system when all the mirrors are in a certain position, because he tried many other positions in which the signal is not that good. From a practical point of view, both Amy and Bob have the same understanding of the laser system, as they can generate the same signal with the same quality. But to whom of the two would we attribute the more sophisticated understanding? To Amy, as she can provide the more sophisticated explanation of why the laser system has to be set up in a certain way to work properly.[18]

Lipton seems to argue for some kind of intuitive or tacit understanding of machines or entities that people can have, like intuitively understanding the engine of one's car or one's computer. *Prima facie*, I agree that such a tacit understanding exists, especially in the context of *practical ends*, but for *epistemic ends* (which is the more common aim for understanding, especially within science) we need another conception of understanding. The reason is that we can assess the appropriateness of

18 Gijsbers is on my side here as well. "Simply knowing how to do something is not the same as understanding how to do that thing (in any significant sense of understanding). This is well known to anyone who has ever followed a step-by-step tutorial for making something work on your computer: even if you learn the tutorial by heart and are able to perform it correctly, that does not mean you understand what you are doing. You may know you have to type "sudo chmod 777 xorg.conf," but you do not understand what you are doing when you type in those signs." Ibid. p. 518.

"practical" or tacit understanding in achieving our practical goals. If it is my goal to fix the engine of my car and I succeed in doing so, one can say that I have some understanding of the engine, as I reached my goal. This kind of understanding, of knowing how to do something or handling an object, tool or instrument, is present in every human domain, including science. However, achieving some practical goal is not the same as achieving the epistemic goal of figuring out what exactly is happening and why, of understanding the behavior of a machinery. From an epistemic point of view, the understanding needs to be made explicit at least to some degree, as otherwise neither the understanding subject nor anyone else could assess whether something of epistemic relevance was understood at all.

In general, and this is my third point of criticism of this example, Lipton's view about causation providing understanding without explanation is based on two problematic assumptions. First, Lipton directly concludes from the assumption that if a person is not able to make all causal information she possesses explicit, but merely some pieces, this explicit information or explanation will not exhaust the understanding. So, whatever it is that she cannot make explicit will be independent from explanation. In other words, he claims that understanding requires or is tight to explanation only if the full understanding can be made explicit through explanation.[19] Lipton's second problematic assumption is that his conception of tacit understanding of causes is compatible with a deep and subtle appreciation of causes. In other words, Lipton is only interested in the difference between someone merely knowing that a phenomenon occurs and someone who has a deep and subtle understanding of the phenomenon. Concerning the first assumption, it is not plausible why understanding should be completely independent of explanation just because no explanation might capture the whole understanding (in this case all the causal knowledge) that a subject has of a phenomenon. Is there any explanation that accommodates this demand? Maybe, but even if an explanation only captures parts of the understanding, *this explanation will be related to the understanding*. And as Lipton himself takes understanding to be a cognitive benefit of an explanation, and not the explanation itself, understanding must somehow be related to explanation. Therefore, the demand that understanding cannot be made fully explicit is not a decisive argument for understanding without explanation at all. Concerning the second problem, Lipton cannot make a convincing case about why tacit understanding can be seen as or identified with deep and subtle understanding of causes, or phenomena more generally. How should that be assessable, for the understanding agent herself or for anyone else? I intuitively agree with Lipton that something like tacit understanding exists and that humans (and possibly other animals) have this tacit understanding. I do think that tacit understanding does not only cover

19 See Lipton (2009), p. 46.

practical understanding, the knowing-how to do something, but can also cover understanding of causes or phenomena. I address this issue in chapter four. However, such a type of tacit understanding should not and cannot be called a deep and subtle appreciation of causes, or a sophisticated understanding of machineries, as there will be no way to determine or to justify whether a subject actually achieved a deep and subtle appreciation of causes without making anything of her understanding explicit. Without providing any explanation, it will not be possible to distinguish a person who has a deep and subtle understanding, and another person who is just lucky in guessing and trying.

In sum, if we accept Lipton's conception of understanding of causes without explanation, we would face an epistemically problematic situation. Taking for granted that understanding is some kind of intellectual or epistemic achievement, the person who wants to understand the why of, say, retrograde motion would want to have access to her understanding. She would want to know whether she understood the causes of a phenomenon correctly, whether she indeed acquired a deep and subtle understanding of the why of retrograde motion. However, according to Lipton's view, she would not have access to her own understanding and could not survey or potentially revise it. Other agents would also never be able to assess whether the subject gained understanding and could never evaluate her understanding as proper or not. The crucial point here is that if a subject cannot provide an explanation, cannot articulate the knowledge or information that she gained, it is unreasonable and impossible to attribute genuine or deep understanding of any phenomenon to that person. There would be no justification at all to attribute genuine or proper understanding to anyone and no ways of identifying potential flaws and improving ones understanding actively and consciously. These are devastating consequences for epistemic endeavors like science, and epistemic achievements in general. Hence, Lipton fails to provide a convincing argument for an understanding of causes that is possible without any relation to explanation, at least for epistemic enterprises like science, enterprises that (primarily) serve epistemic ends. This is not to say that such a kind of tacit understanding does not exist at all, I address this topic in chapter four. What I claim here is that tacit understanding unrelated to explanation is not the kind of understanding that scientists refer to when they mention understanding of phenomena as an aim of science.

### 3.2.2 Necessity

Lipton's second example of cases where understanding is achieved without explanation concerns necessity. He investigates "arguments that are not explanations but do generate understanding by showing necessity."[20] Thought experiments (seem to)

20 Ibid. p. 47.

belong to this kind of arguments and Lipton presents the thought experiment of Galileo as a case in point.

Galileo argued that, according to Aristotelian physics, heavier bodies fall faster to the ground than lighter bodies. Heavier bodies have a higher acceleration. If you stand on top of a tower and let go two masses at the same time, one with a weight of 5 kg and one with the weight of 10 kg, the 10 kg body will reach the ground earlier than the 5 kg body. Following Aristotle, if you tie the two masses together with a rope, the lighter mass should slow down the heavier mass, so that the combined mass will fall slower than the 10 kg body, but faster than the 5 kg body. But this means that a mass of 15 kg (the two masses together) fall slower than a 10 kg mass, which is impossible according to the Aristotelian system. 15 kg cannot fall faster AND slower than 10 kg. Therefore, the assumption that acceleration depends on mass must be rejected.[21]

Imagine someone reads this version of Galileo's thought experiment who did not hear of it before and who has no training in physics, philosophy, or logic. This person then knows the thought experiment in the sense that she can remember it and tell it a third person. But although this person knows the thought experiment, she might not understand it. After reading it, she could ask: So what? What is the point or the problem? My answer could be: The thought experiment shows that *the acceleration of bodies is independent of their mass because it is logically impossible that the acceleration depends on the mass*. The thought experiment shows the logical impossibility. This is an explanation that is included in the thought experiment and that might not be obvious or clear to everyone. The thought experiment as a whole is not an explanation, but it *includes* a logical explanation of why acceleration must be independent of mass. Lipton writes "the system cannot accelerate both slower and faster, so acceleration must be independent of mass."[22] This proposition is an explanation (or at least part of an explanation, depending on the required level of detail concerning the system, acceleration, mass etc.), according to my generic conception, as it provides reasons for why something is the case. The proposition contains an explanans, the logical impossibility of a phenomenon exhibiting contradictory performances simultaneously, and an explanandum, the independence of acceleration of falling bodies from there mass.

Lipton himself addresses the question of why the thought experiment itself should not be regarded as an explanation. His argument is that "Galileo's argument [...] though it gives the necessity and the understanding, seems to me not an explanation. [...] It cannot because the Galilean argument is noncausal, giving no cause of the fact that acceleration is independent of mass. [...] It does not provide

21 See ibid. p. 47. For an English translation of the original thought experiment by Galileo, see Galileo Galilei (1954 [1914, 1638]), *Dialogues concerning two new sciences*. Trans. Crew, H. & de Salvio, A., New York, Dover Publications, pp. 62f.

22 Ibid. p. 47.

a direct answer [to] the question "Why is acceleration independent of mass?""[23] These two features, that the argument is noncausal and does not provide a direct answer to the why-question, are not sufficient to not view parts of the thought experiment as an explanation. First, remember that I argue for an explanatory pluralism that is not limited to causal explanation. The explanation provided by Galileo's thought experiment might be viewed as a logical, counterfactual or a modal explanation, depending on how you conceptualize this type of explanation. In light of the vast amount of literature on non-causal explanation and explanatory pluralism, it becomes even less plausible that knowledge of causes is necessary for understanding.[24] As I already mentioned in section 3.1, the extensive philosophical investigation and literature on different forms of explanation in science show that a pluralist position towards scientific explanation should be adopted. The second feature is a result of Lipton's restriction to causal explanations, as he only takes information about causes to be direct answers to why-questions. While this might often be the case, it is not always so, as in certain contexts, scientific explanations are accepted as direct answers to why-questions although they do not refer to any actual cause. As soon as an explanatory pluralism is adopted and in accordance with my generic conception of explanation, reasons, not only causes, are accepted as direct answers to why-questions.

Additional support for the claim that thought experiments provide understanding through explanation can be found in the literature on thought experiments. For example, James Brown and Ulrich Kühne claim that thought experiments have a crucial function for developing explanations. Both authors argue in favor of the explanatory power of thought experiments throughout scientific history by referring to many other thought experiments in addition to the one from Galileo. Brown explicitly states that Newton wanted to *explain* the existence of absolute space with the bucket (thought) experiment[25] and Kühne argues that a person who accepts the derivation(s) provided by a thought experiment should be able to explain the phenomenon the thought experiment is about. According to Kühne, one function of thought experiments is their use as didactical tools for students who are experiencing a revolution in their personal understanding of nature. One asks for an explanation for a fact *p* if the fact *p* does not fit into the previous understanding of the

23 Ibid. p. 48.

24 For an overview on non-causal explanation and explanatory pluralism, see for example Reutlinger, A. & Saatsi, J. (2018), *Explanation beyond causation: philosophical perspectives on noncausal explanations*. Oxford, Oxford University Press, DOI: 10.1093/oso/9780198777946.001.0001.

25 See Brown, R. J. (1986), "Thought Experiments since the scientific revolution." *International Studies in the Philosophy of Science*, 1 (1), pp. 1–15, DOI: 10.1080/02698598608573279, p. 8. For more details concerning Brown's Platonism, the view that we are able to recognize natural laws *a priori* through the use of thought experiments, see Brown, J. R. (1991), *The Laboratory of the Mind – Thought Experiments in the Natural Sciences*. New York and London, Routledge.

world. One asks "why *p*?" to get an explanation which removes the irregular character of the fact *p* by establishing an acceptable connection of the appearance of *p* with what one regards as the normal course of things. The assertion that we are entitled to consider a factual assertion *p* to be explained if it has been obtained by an acceptable thought experiment is based on this common sense understanding of a good expla nation, so Kühne argues.[26] Kühne's "common sense understanding of explanation" conforms to my generic conception of explanation, as the explanation embedded in the thought experiment provide reasons for *p*, or for why or how *p*.

Again, as in the case of causation, Lipton's view of explanation is much too narrow and he would have to exclude non-causal types of explanations from the realm of explanation, which are nevertheless successfully used and referred to as explanations in scientific practice as well as in the philosophical literature. Thus, there is no convincing reason to assume that thought experiments, or cases of necessity, provide understanding without explanation.

### 3.2.3 Possibility

Then Lipton turns to possibility. Recall that he views understanding to be a benefit of explanation, e.g. the possession of causal information or the apprehension of necessity. In this third case, actual understanding is gained from merely potential explanations, explanations of potential phenomena. Modal understanding is achieved, as in the case of necessity. "Information about other worlds illuminates the actual world. The fact that my computer would not have overheated if the cooling fan had not broken helps to explain why my computer overheated".[27] But this, in fact, is a counterfactual explanation of the breaking of the computer.

Lipton claims that such cases lead to understanding without explanation by arguing that counterfactual explanations, as in the computer example, have a different explanandum than 'real' explanation. In this example, the explanandum of the counterfactual explanation is a *possible* phenomenon, the non-overheating of the computer, and not the *actual* phenomenon, the over-heating of the computer.[28] So, we gain understanding of the phenomenon without an explanation of that phenomenon. If this is actually the case, and if Lipton still wants to allow for the possibility that the counterfactual explanation provides understanding of the possible phenomenon *as well as* of the actual phenomenon, then the counterfactual explanation has to be connected to the actual phenomenon. A subject will have to infer from the understanding of the possible phenomenon (the non-overheating of

26 See Kühne, U. (1997), "Gedankenexperiment und Erklärung." *Bremer Philosophica*, 5, pp. 1–51, pp. 15, 23, 26.

27 Lipton (2009), p. 50.

28 See ibid. pp. 49–52.

the computer) to the understanding of the actual phenomenon (the overheating of the computer). If this inference is not made, the potential explanation would not be connected to the understanding of the actual phenomenon. And this connection can be established by reintegrating the explanation. Instead of saying that the computer would not have overheated if the cooling fan had not broken, one can say that the computer broke because the cooling fan broke. If understanding of actual phenomena shall be possible through potential explanations, which is what Lipton is arguing for, the reason for the actual phenomenon (the over-heating of the computer), the broken cooling fan, must be identified. This again is in line with my generic conception of explanation, which demands that an explanation must provide reasons for the phenomenon. So there is an explanation involved in the understanding of possibilities. If I know the potential explanation that my computer would not have overheated if the cooling fan had not broken, while being unable to make the inference that my computer (probably) broke because of the broken cooling fan, I would not possess modal knowledge in this case, and hence no understanding.

Apart from that, the case that Lipton describes here is completely consistent with Woodward's counterfactual theory of causal explanation. In order for a genuine explanation to count as such, the explanation must provide answers to what-if-things-had-been-different questions. An explanation must exhibit systematic patterns of counterfactual dependence. To put it in Woodward's own words, "to causally explain a phenomenon is to provide information about the factors on which it depends and to exhibit how it depends on those factors. This is exactly what the provision of counterfactual information of the sort described [...] accomplishes: we see what factors some explanandum *M* depends on (and how it depends on those factors) when we have identified one or more variables *S* such that changes in these [...] are associated with changes in *M*."[29] Only by knowing which factor has an effect on a certain phenomenon and how a factor affects the phenomenon is it possible to understand the causal dependence, which is provided by counterfactual explanations.

Another approach that brings Lipton's analysis of this case into trouble is van Fraassen's pragmatic accounts of explanation. If a counterfactual explanation is used to explain an actual phenomenon, and the explanation is in accordance with the respective context in the sense that it provides an answer to a why-question posed, then the actual phenomenon will be explained by the counterfactual explanation. Whether the explanation that the computer would not have overheated if the cooling fan had not broken is evaluated as an adequate answer to the question

29 Woodward, J. (2003), *Making things happen: a theory of causal explanation*. New York, Oxford University Press, DOI: 10.1093/0195155270.001.0001, p. 204. For more details of Woodward's theory, see ibid.

why the computer actually overheated is contextually determined.[30] Van Fraassen argues that "the discussion of explanation went wrong at the very beginning when explanation was conceived of as a relation like description: a relation between a theory and a fact. Really, it is a three-term relation between theory, fact, and context. [...] Being an explanation is essentially relative, for an explanation is an *answer*. [...] It is evaluated vis-à-vis a question, which is a request for information. But exactly what is requested [...] differs from context to context."[31] If such a pragmatic account of explanation is adopted, it can not only accommodate this case of possibility, but also explain why Galileo's thought experiment in the previous example of necessity provides an explanation. If a questioner asks why acceleration must be independent of mass, and receives the answer that acceleration must be independent of mass because the alternative, that acceleration does depend on mass, is logically impossible, and is satisfied with this answer because it fits into his background knowledge, this answer qualifies as an explanation of the acceleration of material objects for this specific questioner.

Hence, Lipton also failed in his attempt to show how understanding in the case of possibility can be acquired without explanation. Lipton's depiction of the case is at odds both with Woodward's counterfactual theory of causal explanation as well as with van Fraassen's pragmatic theory of explanation. In light of both these accounts, it is really implausible that an explanation of a *possible* phenomenon should not be regarded as an explanation that amounts or contributes to understanding the *actual* phenomenon.

### 3.2.4 Unification

The final example Lipton offers to argue for a kind of understanding without explanation is unification. He states that one way science improves our understanding of the world is by showing how diverse phenomena can share underlying similarities. The concrete example of unification as achievement without explanation he presents is Kuhn's account of the dynamics of normal science. From this, Lipton concludes that understanding through unification without explanation is ubiquitous in science. The central question that arises for Lipton is how rule-like behavior can be explained if rules are completely absent. The answer is that this behavior can be explained by shared exemplars. Normal scientists go on to choose new problems that seem similar to the exemplar. Exemplars perform the same function as shared rules, but in contrast to rules, exemplars provide knowledge in an implicit way. The

30 See van Fraassen (1980), pp. 134–157. For more details of van Fraassen's pragmatic account of explanation, see ibid.

31 Ibid. p. 156.

exemplar-based mechanism as proposed by Kuhn is an account of the ability of scientists to select problems that are similar to the exemplar, to try to find a solution for the chosen problem that is similar to a solution of the exemplar, and to assess the suitability of the proposed solution by reference to standards that are upheld for the exemplar. Kuhn mentions the inclined plane, the conical pendulum, Keplerian orbits, and also instruments like the calorimeter or the Wheatstone bridge as examples of exemplars in physics.[32] Lipton argues that the exemplar mechanism provides a plausible example of a route to understanding, i.e. knowledge of unification, without explanation. In an unarticulated way, exemplars provide information about the structure of the world, thereby unify phenomena, and provide understanding by analogy.[33]

Although I do not want to deny that exemplars can play an important role in achieving understanding, I do not think that they can do so without explanation. What Kuhn and Lipton are describing here are skills or abilities. Scientists acquire the skills to choose new problems and work with exemplars by participating in the scientific practice of their community. Through investigating a new problem by reference to an exemplar, scientists will gain new knowledge. In Lipton's words, "one of the ways science improves our understanding of the world is by showing how diverse phenomena can share underlying similarities."[34] Lipton as well as Kuhn are completely right in arguing that the discovery or identification of yet unknown phenomena that share underlying similarities with an exemplar is a matter of skill, not of an explicit theory or explanation. Merely knowing a theory or explanation will not automatically lead to identifying new phenomena. However, identifying or grasping similarity relations between an exemplar and a new problem without any reference to an explanation provided by the exemplar is not possible. When scientists grasp similarity relations, they relate knowledge they already have about the exemplar to the phenomenon that is actually investigated. Kuhn himself states that "learning [from problems to see situations as like each other] comes as one is given words together with concrete examples of how they function in use; nature and words are learned together."[35]

---

32 See Kuhn, T. S. (2012 [1970]), *The structure of scientific revolutions* (4. ed., 50th anniversary ed.). Chicago, University of Chicago Press, p. 186.

33 See Lipton (2009), pp. 52ff. For more details concerning Kuhn's conception of shared exemplars, see Kuhn (1970), pp. 181–190.

34 Ibid. p. 52.

35 Kuhn (2012), p. 190. Kuhn puts so much emphasis on scientific practice, because philosophy of science in his time was almost exclusively concerned with scientific statements, e.g. theories, and their relation to empirical evidence, and regarded actual research practice as uninteresting for philosophical investigation. Although Kuhn's new focus marked the beginning of a crucial change in philosophy of science, later known as the Practice Turn, he never im-

The crucial point is that, according to Kuhn, exemplars are only one component of a disciplinary matrix. Kuhn introduces the concept "disciplinary matrix" in the postscript to *The Structure of Scientific Revolution* to replace and specify his notion of "paradigm". A disciplinary matrix is shared by all members of a particular scientific community, accounts for the functioning communication as well as for the consensus concerning judgements among the professionals, and has four components: symbolic generalizations, shared commitments, values, and exemplars. Exemplars in isolation will not provide, or enable scientists to generate, problem solutions. The four components of the disciplinary matrix are interrelated.[36] For example, group commitments "help to determine what will be accepted as an explanation and as a puzzle-solution."[37] That is, the solutions that scientists find are explicit explanations of phenomena. This becomes apparent in Kuhn's discussion of the impact of Newton's theory on seventeenth century physics:

> Before Newton was born the "new science" of the century had at last succeeded in rejecting Aristotelian and scholastic *explanations* expressed in terms of the essences of material bodies. [...] Henceforth the entire flux of sensory appearances, including color, taste, and even weight, was to be *explained* in terms of the size, shape, position, and motion of the elementary corpuscles of base matter. [...] In an earlier period *explanations* in terms of occult qualities had been an integral part of productive scientific work. Nevertheless, the seventeenth century's new commitment to mechanico-corpuscular *explanation* proved immensely fruitful for a number of sciences, ridding them of problems that had defied generally accepted solution and suggesting others to replace them. [...] The search for a mechanical *explanation* of gravity was one of the most challenging problems for those who accepted the *Principia* as paradigm.[38]

Hence, in Kuhn's account of science, explanations are the problem solutions created by scientists, or at least explanations play an indispensable role in science in order to find solutions. If scientists discover an analogy between the exemplar and a novel phenomenon, they create an explanation, potentially an unificationist explanation, or they employ the same explanans for the exemplar as well as for the phenomenon, for two different explananda. For Kuhn, the process amounts to extending the explanation of the exemplar to the explanation of the new case. Therefore, Lipton is wrong in claiming that Kuhn's account of normal science is a case of understanding without explanation.

---

plied that explicit components of science like theories or explanations were not required to conduct science, and thereby to understand the natural world.

36 See ibid. pp. 181–186.

37 Ibid. p. 183.

38 Ibid. pp. 104f, my emphasis.

The problem that Lipton faces in this example of unification is similar to the problem I point out in his example of causation. In the same way in which explicit knowledge of retrograde motion does not automatically amount to understanding of retrograde motion, merely knowing the explicit content of a theory that covers an exemplar does not automatically allow for an understanding of a new problem. But neither does the mere know-how of how to work with an exemplar without any explicit reference to the theory or established background knowledge. This explicit reference is made in form of an explanation. As in all the other examples, Lipton wants to present cases where "the routes to understanding […] do not pass through explanation."[39] Hence, also in the case of unification, he has the goal to present a case of understanding that is acquired without an explanation coming in at any point in the process of understanding. While Lipton does not mention the concept of tacit understanding again in the case of unification as he did in the case of causation, he nevertheless seems to imply a similar or the same concept here, namely that scientists can link several phenomena through similarity, and not causal, relations. I agree that the processes of choosing problems that seem similar to the exemplar and trying to find solutions that are similar to those that work in the exemplar require, at least partially, tacit processes or skills, a kind of knowing-how. However, I do not see how it should be possible for scientists to generate solutions and judge the adequacy of these solutions in reference to the exemplar without some reference to an explanation. As it is possible that scientists occasionally find solutions that are not adequate according to the standards the respective exemplar represents, no one would ascribe understanding to them in such cases, not even they themselves. As Lipton follows Kuhn in requiring that the scientists should not only be able to choose new problems that are similar to the exemplar and to find solutions for them, but also to assess the appropriateness of the developed solutions,[40] it is not clear how scientists would be able to do that if they cannot provide the solution of the problem in an explicit form, i.e. as an explanation of the new phenomenon that is based on the exemplar.

Concluding, Lipton fails to make a convincing point that understanding of new phenomena through exemplars is possible without any connection to an unificationist or any other kind of explanation. A closer look at Kuhn's account of normal science shows that Kuhn did not separate the process of finding a solution to a problem from explanation, as Lipton wrongly claims. Additionally, as in the case of causation where Lipton discusses understanding through visual representations or handling a machinery, it remains unclear how the appropriateness of a proposed problem solution should be assessed if understanding is a completely tacit process that cannot be made accessible to at least some degree.

39 Lipton (2009), p. 44.

40 See ibid. p. 53.

### 3.2.5 None of these examples is a case of understanding without explanation

In sum, none of the four examples presented by Lipton can reasonably be viewed an instance of scientific understanding without explanation. The reason for this is twofold. In the cases of necessity and possibility, Lipton only accepts a very limited and restricted notion of explanation. He only considers explicit causal explanations that refer to actual causes to count as explanations. This restriction is unreasonable in light of scientific practice, where various kinds of explanations (unificationist, counterfactual, analogue, probabilistic, logical, and this list is not meant to be extensive) are used to achieve understanding of phenomena, and of the various accounts of explanation proposed by philosophers to accommodate the diversity of explanations. In the cases of causation and unification, however, Lipton focuses too much on skills or tacit understanding, which leads to a view of understanding that is too narrow to accommodate the demands that Lipton himself sets up for understanding. He wants that the understanding is assessable, that the understanding subject herself or other agents in the community can judge the acquired understanding as adequate, deep, subtle, or insufficient. However, Lipton does not explain or specify how this should be possible if understanding is tacit and unrelated to any explicit representation like explanation. While I think that Lipton is right in putting so much emphasis on skills or tacit understanding to highlight the difference between understanding a phenomenon and knowledge of a phenomenon, which will be the topic of chapter four, I do not see how this claim automatically qualifies or amounts to understanding being *completely* separated from explanation. Especially when we think about epistemic activities like science and scientific understanding, it remains unclear in which way such a form of tacit understanding would be more valuable than understanding that can (partially) be made explicit and hence evaluated. And as I argue in this section, Lipton could not give a convincing argument to this effect.[41]

Independent of the persuasiveness of Lipton's position and his examples, the question about the relation of explanation and understanding has not lost any of its topicality in the last decade. In this context, two kinds of understanding lie at the center of the discussion: objectual and explanatory understanding. The former is said to be possible without explanation, whereas the latter, as the name suggests, is based on explanation. I now turn to these two kinds of understanding and the debate that emerged around them.

---

41 For another line of argument why Lipton's examples as cases of understanding without explanation fail, see Khalifa (2017b), chapter 5.

## 3.3 Objectual and explanatory understanding - a controversy

Lipton's work did not settle the question of whether understanding could be possible without explanation. Quite the contrary, the (possible) relation to explanation still is a central topic in the debate on understanding. This concerns philosophers of science as well as epistemologists. In the same year in which Lipton's cases for understanding without explanation were published, Jonathan Kvanvig introduced a different argument for the possible independence of understanding from explanation, which is not related to Lipton's examples at all, and a new terminology that should become formative for the debate on understanding, namely objectual and explanatory understanding. In section 3.3.1, I present Kvanvig's notions of objectual and explanatory understanding and his argument for their difference in (not) being related to explanation. Kvanvig's argument is extensively addressed by Kareem Khalifa and section 3.3.2 is devoted to Khalifa's criticism of Kvanvig's distinction. In section 3.3.3., I relate Khalifa's critique on Kvanvig's argument to my criticism of Lipton's view of understanding without explanation. I conclude that Lipton and Kvanvig, while presenting different and independent arguments for a separation of understanding and explanation, both make the same mistake of having a too narrow view of (scientific) explanation that is not defendable in light of an explanatory pluralism, which is supported by scientific practice and the various philosophical accounts of explanation. Thus, also Kvanvig fails to provide an argument for why scientific understanding should or could be independent of explanation.

### 3.3.1 Kvanvig's argument for a distinction of objectual and explanatory understanding

In order to make sense of Kvanvig's distinction between objectual and explanatory understanding, I will first lay out some claims he makes about understanding in general. At the beginning of his analysis, Kvanvig points to the different foci that investigations on knowledge or understanding have. When investigating knowledge, the focus lies on issues like what the evidence of a belief is, how reliable a belief is, or whether the connection between the reasons for a certain belief and the truth of this belief were formed accidentally. When understanding is the target of investigation, other questions are addressed. How are pieces of information connected to each other? What is the extent of the grasp of structural relations between the central items of information regarding which the question of understanding arises? Concerning understanding, questions about structural relations between pieces of information that are grasped arise and are addressed, while investigations on knowledge focus on questions like the non-accidentality or justification of knowledge.[42] So, in contrast to Lipton, Kvanvig argues that understanding is not reducible to knowledge. This differentiation will be important for the chapters to come, though not for the discussion in this chapter.

In the case of understanding, the body of information that an individual possesses is constituted by a grasped relatedness of pieces of information. Importantly, Kvanvig claims that the mere existence of explanatory and other connections between these items or the easy accessibility of these connections are not enough for understanding. An already-possessed awareness of the connections is also required. An already-mastered grasp is needed to recognize the connections. If this grasp is absent, there can be as many obvious relations between pieces of information as you want, they would not be recognized by a subject and, therefore, the subject would not understand the body of information. In short, Kvanvig characterizes understanding as grasping structural relations and grasping amounts to making sense of the object of understanding.[43] Unfortunately, Kvanvig does not clarify the notion of grasping further and it remains obscure what it means that a subject is able to make sense of an object. I will return to the concept of grasping in section 4.3.1.

According to Kvanvig, the structural relations that can be grasped by a subject include not only explanatory, but also logical and probabilistic relations, and explanatory relations are only incorporated into understanding when they exist. A subject can have objectual understanding of an indeterministic system by grasping logical or probabilistic relations present in this system even if no explanatory relations between parts of the system exist. He uses an example to illustrate this point. The reader is asked to imagine an electron that goes to the left rather than to the right. The probability of the electron going left is exactly the same as the probability of going to the right. Such a quantum-mechanical system is an indeterministic system, we will not know in advance which way the electron will take. Kvanvig claims that whichever way the electron will go, it will do so by chance and there is no cause of why the electron goes that way. "If there is no cause of the electron going to the left rather than the right, there is no explanation why the electron went to the left either."[44] According to Kvanvig, it is possible to objectually understand such indeterministic systems by grasping logical or probabilistic relations, but it is not possible to have explanatory understanding of such a system, because there are no explanatory relations between facts of the system. In short, Kvanvig argues that we are able to objectually understand indeterministic systems that cannot be explained because

---

42 See Kvanvig, J. L. (2009), "The value of understanding." In Haddock, A., Miller, A. & Pritchard, D. (eds.), *Epistemic value*, pp. 95–111, Oxford, Oxford University Press, DOI: 10.1093/acprof:oso/9780199231188.001.0001, pp. 96f.

43 See ibid. pp. 97ff.

44 Ibid. p. 101.

they do not contain explanatory relations. However, such systems contain probabilistic or logical relations that can be grasped. Therefore, objectual understanding does not reduce to explanatory understanding, which is a type of propositional understanding.[45]

After presenting the example of the electron, Kvanvig concludes that objectual understanding cannot be reduced to explanatory understanding, because there is no causal explanation or relation of the event that could be grasped, then possessed by and attributed to a subject. It is not possible to state that Jill understands why the electron went left, since she could not grasp any explanatory relation. Notwithstanding this differentiation, Kvanvig claims that a unified conception of understanding, in contrast to the concept of knowledge, should be aspired. Neither objectual nor explanatory understanding are reducible to knowledge. We can objectually understand indeterministic systems and we can explanatorily understand deterministic systems, since deterministic systems contain explanatory relations that we can grasp. In both cases, understanding amounts to grasping structural relations, which is something different than having knowledge. Kvanvig does think that understanding as well as knowledge can be subdivided into their propositional, explanatory and objectual forms, but these do not affect the general difference between understanding and knowledge. In all cases of propositional, explanatory or objectual knowledge, something like non-accidentality is of interest, whereas cases of objectually understanding indeterministic systems and explanatorily or propositionally understanding deterministic systems comprise a grasp, a sense-making, of the relations involved, which is not covered by any of the forms of the concept 'knowledge'.[46]

To summarize, Kvanvig argues that understanding is the grasp of structural relations of the object that should be understood. A subject gains explanatory understanding of the object if she grasps explanatory relations, and she gains objectual understanding if she grasps logical, probabilistic or any other kind of relations that

45 See ibid. pp. 101f. Kvanvig merely refers to an intuition that "it is tempting to adopt the thesis that [explanatory understanding] can be explained in terms of propositional understanding." Ibid. p. 96. However, the identification of explanatory understanding with propositional understanding is a debated issue. For example, Christoph Baumberger argues that a reduction of "explanatory to propositional understanding is either impossible or unhelpful." Baumberger (2011), p. 87. He sticks to this opinion and provides the same argument in his later work again, namely "that explanatory understanding cannot be reduced to propositional understanding." Baumberger, Beisbart & Brun (2017), p. 25.

46 See ibid. pp. 97, 101f. Kvanvig then goes on to argue that understanding is compatible with Gettier-cases or types of epistemic luck, which is not the case for knowledge. This observation provides additional support for the distinction between understanding and knowledge, see ibid. pp. 103–109. Since I am concerned with the relation of understanding and explanation in this chapter, and not with the relation of understanding and knowledge, I will not discuss the question of the possible compatibility of understanding with epistemic luck.

are not explanatory. The concept 'understanding' always refers to (the extent of) this grasp, while the concept 'knowledge' addresses issues of reliability or non-accidentality of beliefs. Therefore, understanding in all its forms is not reducible to any form of knowledge.

### 3.3.2 Khalifa's argument for a reduction of objectual to explanatory understanding

Although I agree with Kvanvig that understanding cannot be reduced to knowledge, a claim I elaborate in chapter four, I disagree that objectual understanding and explanatory understanding can be clearly distinct. Kareem Khalifa is not convinced by Kvanvig's argumentation either and directly addresses Kvanvig's account. I now present Khalifa's arguments against Kvanvig, before I relate Khalifa's criticism of Kvanvig's distinction to my arguments against Lipton's examples in the next section.

Khalifa identifies four features that Kvanvig seems to assume in the system of the moving electron, which he provides as an example for objectual understanding without explanation:

1. The explanation has to be *causal*: "if there is no cause of the electron going left rather than right, there is no explanation why the electron went to the left either."

2. The explanandum is *indeterministic*: "In indeterministic systems, things happen that are uncaused, both probabilistically and deterministically."

3. The explanandum is *contrastive*: "the events in question are irreducibly indeterministic in such a way that there is no causal explanation as to why the actual events occurred rather than some other events."

4. The explanandum contrasts *equally probable* outcomes: "If the probability of an electron going to the left is precisely the same as that of going to the right (and there is no hidden variable to account for the difference), then whichever way it goes is the result of chance rather than causation."[47]

Khalifa concludes that "Kvanvig is denying the possibility of causal, indeterministic explanations of explananda contrasting equally probable outcomes."[48] Khalifa addresses all of the four features in turn to show that there are in fact explanatory relations present in the electron-example, which implies that it is possible to explana-

47 Khalifa, K. (2013), "Is understanding explanatory or objectual?" *Synthese*, 190, pp. 1153–1171, p. 1158, DOI: 10.1007/s11229-011-9886-8, original emphasis.

48 Ibid. p. 1158.

torily understand indeterministic systems and amounts to a reduction of objectual to explanatory understanding.

First, there is no good reason to limit the notion of explanation to causal explanation. Khalifa also argues for an explanatory pluralism. The mere fact that several kinds of explanations exist (causal, deductive, model-based, unificationist, mechanistic, functional, probabilistic, counterfactual, among others) and are used in scientific practice is a good reason to take all these kinds of explanation to be permissible in certain contexts and all of these kinds can provide explanatory understanding. If scientists in Kvanvig's example grasp the logical or probabilistic relations that he takes to be present, they will perform some reasoning about the system considering the "probability distributions about an electron's position [derived] from its quantum state. [To do so, scientists will have to incorporate the set of] quantum numbers and the eigenfunction that characterize the possible states of the quantum mechanical system"[49]. This line of reasoning can definitely be characterized as an explanation based on a theory, in this case, quantum mechanics, according to Khalifa.[50]

Second, Kvanvig identifies indeterminism with the absence of causes. However, as Khalifa highlights, the fact that a system is indeterministic does not automatically mean that there are no causes in play. It simply means that the same cause can produce varying outcomes. Therefore, there might causal relations (and hence, explanatory relations on Kvanvig's account). Moreover, if one accepts the first critique and does not limit the notion of explanation to causal explanation, it becomes even more obvious that we can have explanations of indeterministic systems. Some theories of explanation admit of indeterministic explanations. Christopher Hitchcock presents the core idea of indeterministic explanations as "a factor *A* is explanatorily relevant to [an explanandum] *E* if *A* plays a non-eliminable role in determining the probability of *E*."[51] If Kvanvig denies the possibility of explanations of this form, he is at odds with scientific practice where indeterministic explanations of the type conceptualized by such philosophical accounts can be found, so Khalifa. Since indeterministic explanations include theoretical statements, scientists derive probabilities or chances that Kvanvig views as non-explanatory from theories that are undeniably explanatory.[52] By using quantum theory, scientists can explain "why an electron had

49 Ibid. p. 1158.

50 See ibid. p. 1158.

51 Ibid. p. 1159. For more details concerning Hitchcock's argument, see Hitchcock, C. R. (1999), "Contrastive explanation and the demons of determinism." *British Journal for the Philosophy of Science*, 50 (4), pp. 585–612, DOI: 10.1093/bjps/50.4.585. For further information about probabilistic explanation, see e.g. Railton, P. (1978), "A Deductive-Nomological Model of Probabilistic Explanation." *Philosophy of Science*, 45 (2), pp. 206–226, DOI: 10.1086/288797; or Strevens, M. (2008), *Depth*. Cambridge (MA) and London, Harvard University Press.

52 See ibid. p. 1160.

a probability *p* of being in a spatial region *x* at a given time interval *t*. [...] If the quantum state were different, then the probability of the electron being in a spatial region (e.g. "the left") would be different."[53]

So, it is possible to indeterministically explain why the electron went left. But what about the third requirement, that the explanandum is contrastive? If it is not possible to explain "why the electron went left *rather* than right", will this strengthen Kvanvig's argument? In other words, do contrastive explanations imply determinism? Not necessarily, according to Khalifa. He draws on Glymour's notion of parity, which states that all possible outcomes of a system can be explained by using the same information.[54] The same information (namely the derivations from the electron's quantum state) do explain why the electron went left, why it did not go right, why it could have gone right etc. Because the system is indeterministic, no further information are relevant for the contrast. In fact, there is no contrast in an indeterministic system that could be grasped, neither explanatorily nor objectually, "because the same factors produce both a likely outcome and an unlikely one – that is the crux of indeterminism."[55] And everything that is close to the contrast (e.g. why the electron did not go right) will be explained by an indeterministic explanation. The two explananda ("the electron went left" and "the electron did not go right") have the same explanans, namely the respective quantum states of the electron. Therefore, parity supports the reduction of objectual understanding to explanatory understanding.[56]

However, Kvanvig could object that with this strategy of parity, we are actually explaining different explananda than the one he offers in his example. Drawing on parity, we can explain "why the electron went left", "why it did not go right", or "why it could have gone right", but we do not explain the contrastive expanandum "why the electron went left *rather* than right". Fortunately, Hitchcock's account of contrastive indeterministic explanations provides a solution to this objection.[57] "*A* [should be viewed] as explanatorily relevant to the contrastive question 'why *E* rather than *F*' if *A* continues to be relevant to *E* when the (exclusive) disjunction *E* v *F* is held fixed. [...] This means that *A* is explanatorily relevant to *E rather than F* when $P(E|(A\&B)\ \&\ (E \lor F)) \neq P(E|B\ \&\ (E \lor F))$."[58] *B* represents the given background conditions that are held fixed. Let's consider a pedestrian example. The explanandum that shall be explained is "Mary ate candy rather than fruits on Friday" (*E* v *F*), although she is on a diet

53 Ibid. p. 1161.

54 See Glymour, B. (2007), "In defence of explanatory deductivism." In Campbell, J. K., O'Rourke, M. & Silverstein, H. (eds.), *Causation and explanation*, pp. 133–154, Cambridge (MA), MIT Press.

55 Khalifa (2013), p. 1161.

56 See ibid. pp. 1161f.

57 See ibid. pp. 1161f.

58 Hitchcock (1999), p. 587.

for weeks (*B*). The explanatorily important factor *A* is that Mary's two best friends, who she met on Friday, offered her some candy. The explanans "because her friends offered her candy" explains both explananda "Mary ate candy" as well as "Mary ate candy rather than fruits". If Mary had not met her friends, the probability of her eating candy would have been much lower. In either case, the structural relations that could be grasped can be incorporated into an explanation of the system.

The last feature of Kvanvig's example is the equal probability feature of the system. One could deny that contrastive explanations imply determinism and then argue that contrastive indeterministic explanations are possible only when the probabilities of the outcomes are different. In this case, since the probabilities of the electron's going left or right are the same, the event could not be explained, and some kind of non-explanatory understanding is involved. This position is unreasonable, according to Khalifa, since it confuses the source of explanatory relevance. It is not necessary for an explanation to make the probabilities of two events different from each other. It can be explained why there is the equal probability of 50% to get head or tail when tossing a fair coin, namely because it is a fair coin with only two realizable options. Hence, an explanation might allow for identical probabilities of two events. Only if the explanans be different would the explanation have to account for these differences in probability. For example, the probability of a coin showing head is 70%, because it is not a fair coin. Additionally, Hitchcock's account of contrastive indeterministic explanations accommodates cases of identical probabilities of outcomes.[59] In sum, there is nothing special about equally probable outcomes.[60]

Khalifa concludes that there is nothing that "precludes the possibility of indeterministically explaining a contrast between two equally probable outcomes"[61]. He calls his general objection to Kvanvig's position and his example *the hidden explanation objection*, according to which logical or probabilistic relations can be explanatory.[62] Kvanvig's restriction of explanations to causal explanations that require the presence of causal relations between events is based on an assumption that is not reasonable in light of an explanatory pluralist position. Probabilistic and logical relations do figure into explanation, and thereby into understanding as well, in a variety of ways: at least, they can play an explanatory role. Khalifa calls this the *Explanatory Role Assumption*.[63] He offers four examples for this assumption:

---

59 That is, $P(E|(A\&B)\ \&\ (E \vee F)) \neq P(E|B\&(E \vee F))$ is consistent with equally probable outcomes, i.e. $P(E|(A\&B)\ \&\ (E \vee F)) = P(F|(A\&B)\ \&\ (E \vee F))$.

60 See Khalifa (2013), pp. 1162f.

61 Ibid. p. 1163.

62 See ibid. p. 1157.

63 See ibid. p. 1165.

> 1. Logical and probabilistic relationships are frequently explanatory, [as in the cases of indeterministic and contrastive explanations].
>
> 2. Logical and probabilistic information may be either an explanans or an explanandum. [...] Since explanantia and explananda are essential elements of an explanation, [they can be incorporated into an explanation.]
>
> 3. Even when logical or probabilistic relationships are not "directly" explanatory, they may still be justifying parts of an explanation (i.e. the explanans, explanandum, or the fact that the two stand in a given explanatory relation).
>
> 4. Logical and probabilistic relations can facilitate correct explanations by specifying the presuppositions of a correct explanation.[64]

The first example, where logical and probabilistic relations are directly explanatory, is demonstrated by Khalifa's objection to Kvanvig's example of the moving electron, in which the event (explanandum) is explained in terms of the present probabilistic relations (explanans). That is, the electron went left because it had a certain probability to do so due to its initial quantum state. The second example is supposed to highlight that probabilistic or logical relations can also figure in an explanation if they are the explanandum, and not the explanans, as in the first case. The probability of a coin showing head is 50%, because a fair coin has only two sides that can show up and none of the two sides is favored.

In the third example, situations are addressed in which grasped logical or probabilistic relations give a better justification for an explanation so that the goodness of an explanation improves. In these cases, the explanatory understanding of a subject improves, but she will not have an additional, irreducible form of objectual understanding. Unfortunately, Khalifa does not provide an intelligible example for this third claim. If I understand Khalifa correctly, one could say that if I understand the stability of atoms through the features presented by Bohr's model of the atom, the information that Bohr's model of the atom also explains the emission of spectral lines described by the Rydberg formula provides additional justification for my understanding of the stability of atoms in terms of Bohr's model. However, I admit that this third explanatory role that Khalifa ascribes to logical or probabilistic relations is the least comprehensible one.

In the final example, by specifying presuppositions of an explanation, logical and probabilistic relations add aspects of an explanation.[65] "For instance, my arm bumping the inkwell explains why it spilled, and the inkwell's spilling presupposes (e.g., through logical entailment and auxiliary assumptions) that an object is extended in

64 Ibid. pp. 1165f.

65 See ibid. pp. 1165f.

space. The relationship between the inkwell's spilling and the presupposition is not explanatory, yet without the presupposition, correctly explaining it would be difficult if not impossible (e.g. try explaining why the inkwell spilled if it could have been a one-dimensional object)."[66] The crucial thing to note here is that not every relation or information that contributes in some way to an explanation must itself be explanatory or even explained. In the case of the spilled inkwell, the presupposition that inkwell is extended in space on its own is not explanatory at all for why the inkwell spilled. And furthermore, it is also not necessary to have an explanation of why objects are extended in space. Logical or probabilistic relations can be indispensable for explaining a specific event or object, but they themselves neither need to be explanatory nor explained.[67]

In short, Khalifa argues that objectual understanding, at least in the form in which Kvanvig introduced it, is reducible to explanatory understanding, because the presence of logical or probabilistic relations provides understanding only if they play one of the four explanatory roles.[68] If probabilistic or logical relations do not figure in an explanation by taking one of the four roles, they will not be incorporated into understanding, since it would not be clear how they are related to the phenomenon that should be understood. Therefore, understanding always requires some form of explanation.

### 3.3.3 The flaws of separating scientific understanding from explanation

Summing up, Kvanvig tries to show with the example of the electron that we can have objectual understanding of a system for which we have no explanation. In his view, we cannot explain indeterministic systems, as these systems do not contain explanatory (causal) relations, but only probabilistic or logical relations. Since understanding is grasping structural relations, we can have explanatory understanding only if we grasp explanatory relations. Thus, we can explanatorily understand deterministic systems, and we can objectually understand indeterministic systems. According to Khalifa, this distinction is not plausible, as scientists can and regularly do explain indeterministic systems, including quantum mechanical systems, through relating probabilistic or logical relations to an explanation.

Is there some common ground in the attempts of Lipton and Kvanvig to specify cases or types of understanding without explanation? Both provide different and independent arguments and examples and do not refer to each other's work. However, a closer look reveals that, despite the differences, both views suffer from the same two flaws. First, they are explicit about only viewing causal explanations to be actual

---

66 Ibid. p. 1166.

67 See ibid. p. 1166.

68 See ibid. p. 1166.

explanations and they do not allow or consider a pluralism of explanation. If an explanatory pluralism and a generic conception of explanation are adopted, for which I argue in section 3.1, the cases provided by Lipton and Kvanvig are explanatory, just not in a causal sense. It will probably be very hard to find cases of understanding that work without any reference to some type of explanation in this generic sense.

Second, Lipton and Kvanvig do not pay sufficient attention to scientific practice. This becomes particularly clear in Kvanvig's example of the moving electron. Contrary to what Kvanvig claims, physicists can and do explain why electrons or other subatomic particles do certain things based on quantum theory, as Khalifa extensively elaborates. And Lipton would have anticipated the problem his view is facing, namely conceptualizing understanding (of phenomena) as something tacit, while demanding that the quality, the adequacy or depth, of understanding (of phenomena) can be accessed and evaluated. If understanding would be completely tacit and unrelated to explanation, or any other explicit representation, it could not be assessed. Quite obviously, both flaws are related. A closer look at science would have shown that narrowing explanation to causal explanation and taking understanding (of researched phenomena) to be (completely) tacit does not accord with scientific practice. I will substantiate this claim and analyze the relation of tacit and explicit dimensions in science based on the work of Michael Polanyi in chapter four. In any case, various types of explanation are ubiquitous in scientific publications and discourse, and the discoveries made in research are grounded in understanding, at least ideally. Scientists explicitly communicate what they discovered and understood about a phenomenon, the results of their research, and they often do so by using explanations, in order to argue why they think that something about the phenomenon is the case, to provide reasons for why they think that a newly discovered insight about a phenomenon is actually the case. How exactly understanding and knowledge are related is the topic of chapter four. The point I want to make here is that, independent of what understanding turns out to be exactly, neither Lipton nor Kvanvig provide a convincing argument for why scientific understanding of researched phenomena should be separated from explanation. This would leave open the question of how the myriad explanations present in science are related to understanding, and what a type of tacit or non-explanatory understanding would add to explanatory understanding acquired in science that explanatory understanding does not already offer.

Nonetheless, the debate about explanatory and objectual understanding still is far from being settled. Both conceptions gained and maintained central importance in the literature on understanding. In the next section, I address and analyze further attempts to justify a distinction between objectual and explanatory understanding, and therefore a separation of understanding from explanation, and argue that these attempts also fail to make such a separation conclusive for scientific understanding.

## 3.4 Further attempts to differentiate objectual and explanatory understanding

Kvanvig and Khalifa are by far not the only authors who are concerned with the concepts of objectual and explanatory understanding. In general, objectual understanding is treated as a broader type of understanding than explanatory understanding. Catherine Elgin, for example, follows Kvanvig and distinguishes between propositional understanding, which involves grasping a fact and covers explanatory understanding, and objectual understanding, which consists of grasping a range of phenomena.[69] Baumberger, Beisbart & Brun introduce the following differentiation between the two types of understanding:

(OU) *S* understands some subject matter or domain of things;
(EU) *S* understands why something is the case.[70]

In this section, I present further arguments in favor of and against a distinction between explanatory and objectual understanding and argue that for scientific understanding of phenomena, the distinction between objectual and explanatory understanding remains untenable, also in light of these further arguments.

### 3.4.1 Differentiating objectual and explanatory understanding according to their targets and vehicles

Christoph Baumberger is another proponent of a distinction between objectual and explanatory understanding. He distinguishes three different types of understanding in terms of the object or target that is understood and the vehicle by which the object is understood. In the case of objectual understanding, the object that is understood is a subject matter, a topic or system (like electromagnetism or global warming), and the vehicle is a comprehensive body of information, like a whole account or theory. The object of explanatory understanding are phenomena that are in some way narrower, like events (the appearance of a certain electromagnetic field or the rise of temperature). The same holds for the vehicle. Whereas a whole body of information or theory is required for objectual understanding, an explanation is the necessary vehicle for explanatory understanding. Finally, Baumberger addresses propositional understanding, by which a fact can be understood (e.g. this particle

69 See Elgin (2017), chapter 3. Interestingly, in an earlier paper Elgin states that objectual understanding is the kind "of understanding that is closely connected to explanation." See Elgin, C. Z. (2007), "Understanding and the Facts." *Philosophical Studies*, 132, pp. 33–42, p. 35, DOI: 10.1007/s11098-006-9054-z.

70 Baumberger, Beisbart, & Brun (2017), p. 5.

is charged, or the temperature has increased) through a proposition.[71] However, he later argues that the concept of propositional understanding is useless, because it either reduces to propositional knowledge or it amounts to explanatory understanding. Therefore, there are only two types of understanding, namely explanatory and objectual understanding.[72]

And these two kinds of understanding are genuinely distinct, according to Baumberger, because explanatory understanding is neither sufficient nor always necessary for objectual understanding. In addition to understanding why some event involved in a subject matter occurred, it also has to be understood what effects the subject matter might have and how it is related to all kinds of other subject matters or systems. Conceiving objectual understanding of a subject matter as explanatory understanding of significant subsets of events concerning the subject matter is not feasible, because explanatory understanding does not include an "awareness of how the different explanations fit into, contribute to, and are justified by reference to a more comprehensive understanding in which they are embedded."[73] These requirements are not fulfilled by Baumberger's conception of explanatory understanding, which is the reason why explanatory understanding is not sufficient for objectual understanding.[74] He provides the following example to illustrate the difference between explanatory and objectual understanding:

> Understanding a subject matter involves more than understanding why some fact involved in it obtains. Besides understanding why it occurs, understanding global warming involves, for instance, understanding what effects (on natural and social systems) it will have, how it is linked to human activities (such as burning fossil fuels and deforestation) and related phenomena (such as the destruction of stratospheric ozone and global dimming), how far greenhouse gas emissions and, as a result, temperatures are likely to rise in the future, and how the changes will vary over the globe. A broader understanding of global warming may even involve instrumental and moral understanding, such as understanding the (dis-)advantages of different responses to climate change (such as mitigation, adaptation and geoengineering), and what a just distribution of emission rights amounts to.[75]

---

71 See Baumberger (2011), p. 71.

72 See ibid. pp. 86f. I do agree with Baumberger that a conception of propositional understanding is not helpful or illuminating in the discussion about the nature of understanding. This is why I am only concerned with explanatory and objecutal understanding, too. However, even if a useful concept of propositional understanding will be developed in the future, it remains to be seen whether this new conception has any effects on the concepts of explanatory and objectual understanding.

73 Ibid. p. 78.

74 See ibid. pp. 77f.

75 Ibid. pp. 77f.

In this example, explanatory understanding of the events involved in the subject matter (e.g. the increase of greenhouse gas emissions due to human activity) is included in, and therefore a part of, objectual understanding, but explanatory understanding of these aspects does not exhaust the objectual understanding of global warming. Hence, explanatory understanding is not sufficient for objectual understanding. And in some cases, objectual understanding of a subject matter can be gained without any relation to explanatory understanding, as the following example shall demonstrate:

> Eighteenth-century biology, conceived of as a pure science of classification with no interest in explanation but with rigorous criteria of success, seems to provide some understanding of the animal kingdom since its classifications reveal significant similarities and allow successful predictions (e.g. about whether an animal of a hitherto unknown species is warm- or cold-blooded). However, this understanding cannot be formulated as understanding why something is the case (e.g. why some organism has a certain feature or why animals of a certain species exist).[76]

In Baumberger's view, these classificatory theories are not the best vehicle to understand animals, exactly because they cannot provide answers to all these why-questions. Evolutionary theory, which did provide these answers, marked a great advance in understanding. However, since understanding comes in degrees and because the classificatory theories provided at least some understanding, as Baumberger claims, it is wrong to think that explanatory understanding is always and in every case included in objectual understanding. Therefore, explanatory understanding is not necessary for objectual understanding. Still, according to Baumberger, both examples show that *typically* explanatory understanding is a part of objectual understanding. In order to have objectual understanding and not merely explanatory understanding, more relations have to be grasped by using a more comprehensive body of information as it would be necessary for explanatory understanding. Baumberger identifies grasping with the manifestation of certain abilities and claims that more of the same abilities, which are already necessary for explanatory understanding, are needed in the case of objectual understanding.[77]

---

76 Ibid. p. 78. Baumberger took this example from Gijsbers, who presents an account of understanding through unification without explanation. However, Gijsbers, as all authors who argue for an account of understanding without explanation, employs a very narrow notion of explanation, since he ties explanation to determination. As I have already argued in sections 3.1 and 3.3, if we allow for an explanatory pluralism that is neither tight to causation nor to determination nor any other concept, we will see that the animal kingdom can be explained in terms of similarity or kinship. For more details concerning Gijsbers' account, see Gijsbers (2013).

77 See ibid. pp. 78f. The concept of grasping will be further discussed in section 4.3.1.

However, Baumberger cannot argue convincingly that explanatory understanding is neither sufficient nor necessary for objectual understanding. Basically, his argument for the claim that explanatory understanding is not *sufficient* for objectual understanding rests on two assumptions. First, objectual understanding requires *more* of the same abilities as explanatory understanding to grasp more relations of various kinds, as explanatory understanding is limited to the understanding of causes of an event, whereas objectual understanding includes probabilistic, teleological, conceptual and other forms of relations as well.[78] This argumentation only shows a difference in degree between explanatory and objectual understanding, but not in kind, precisely because *more of the same* abilities are necessary. If Baumberger could have shown that objectual understanding requires genuinely different abilities than explanatory understanding, his claim would stand. He does not show that, as he states that "compared with a single instance of explanatory understanding, objectual understanding of a subject matter involves grasping more dependence (and similarity) relations in and by means of a more comprehensive body of information [...] Understanding global warming involves more of the same abilities as does understanding its causes."[79] And second, as Lipton and Kvanvig, Baumberger also seems to have a very narrow concept of explanation that only allows for causal explanation, since explanatory understanding amounts to abilities to understand causes of an event. As I especially argued throughout sections 3.1 and 3.3, such a narrow causal notion of explanation is inappropriate for accommodating the diverse types of explanations found and employed in scientific practice. If a pluralistic notion of explanation is accepted, various kinds of explanation can easily accommodate the grasping of various kinds of relations present in a subject matter, which can serve an explanatory purpose or role. If more of the various kinds of relations are grasped than only causal relations, the explanatory understanding will be better.

To show that explanatory understanding is not even *necessary* for objectual understanding, Baumberger presents the example of eighteenth-century biology and claims that these theories provide understanding of the animal kingdom without explaining why animals have specific characteristics. Therefore, one can achieve objectual understanding with these theories as vehicles, but not explanatory understanding. I disagree and claim that these theories do not provide understanding without some kind of explanation. True, classificatory theories do not causally explain why animals have certain characteristics or why they exist at all, but the successful prediction of whether a newly discovered species is warm- or cold-blooded can be explained by referring to the classificatory criteria provided by the theory. Whether a new species is warm-or cold-blooded can be explained by reference to already established taxa and their assigned criteria. It will be predicted that, for example, a

78 See ibid. p. 79

79 Ibid. p. 79.

new turtle species is cold-blooded exactly *because* it belongs to a taxon whose members are all cold-blooded. The newly discovered species must share some characteristics with already established taxa, and based on the identified (dis-)similarities, the membership to a specific taxon can be clarified and additional characteristics predicted. If you discover a new animal species and see that this animal has flakes, a carapace and flippers, you conclude that it is a turtle. Based on that insight, you predict that this animal is also cold-blooded, because turtles are cold-blooded.

In light of some standards, this might not be a very good or satisfying explanation, but it clearly is an explanation. Again, if the notion of explanation is not limited to causal explanation, but instead seen as a representation that provides reasons for features of the phenomenon, like my generic conception of explanation introduced in section 3.1, explanations are involved in eighteenth-century biology as well. Classificatory biology provides reasons for why new discovered reptile or fish species will be cold-blooded, while new discovered birds or animal species will be warm-blooded. If predictions based on a classification system fail, this would point to flaws in your classification system and also in your understanding of the animal kingdom. If you cannot provide reasons for why you made a certain prediction, why you think that a new animal species will have a certain characteristic, you would not have made a prediction, but merely a guess, as you would not have understood the animal kingdom through the classificatory theory. Since Baumberger even goes on to argue that evolutionary theory, which provides explanations, trumped the other biological theories and that, typically, objectual understanding includes explanatory understanding, he does not provide a convincing argument for the independence of objectual understanding from explanatory understanding. Trying to claim that explanatory understanding is not necessary for objectual understanding seems to be an artificial and counter-intuitive move.

In sum, Baumberger argues that objectual and explanatory understanding are distinct, as these types of understanding have different objects as their targets (subject matters, topics, or systems vs. events) and employ different vehicles (whole accounts or theories vs. explanation). They differ in terms of vehicles and targets, because explanatory understanding is neither necessary nor sufficient for objectual understanding. Explanatory understanding is not sufficient for objectual understanding, as objectual understanding requires more of the same abilities, and it is not even necessary for objectual understanding, since it is possible to understand phenomena in the world based on theories that do not provide explanations, like classificatory theories in biology. I have argued in this sections that Baumberger fails to provide a convincing argument for why explanatory understanding is neither sufficient nor necessary for objectual understanding. On the one hand, as he demands that for objectual understanding more of the same abilities as for explanatory understanding are required, the difference between explanatory and objectual understanding will only be a matter of degree, but not in kind. Objectual understanding would just be a label for 'better' or 'more comprehensive' explanatory understanding, but we will always talk about the same kind of understanding, namely explanatory understanding. On the other hand, Baumberger, like Lipton and Kvanvig, limits his notion of explanation to causal explanation. Having explanatory understanding of an event amounts to grasping its cause. Taking explanatory pluralism into account and allowing reasons to be a variety of explanantia, and not only causes, reveals the ubiquitous presence of explanation and their function of relating theory and worldly phenomena. So, the ground on which Baumberger rests his claim that objectual and explanatory understanding differ in terms of their targets and vehicles is collapsed. Or is it not?

### 3.4.2 Why a differentiation in terms of the target does not hold

In the work of Stephen Grimm, I find additional support for the identification of objectual and explanatory understanding, since Grimm does not think that the distinction between the two forms should or could be drawn, either. Again, Baumberger differentiates between explanatory and objectual understanding in terms of their vehicles (explanation vs. theory) and their objects (event vs. whole subject matter). Grimm agrees that the objects of understanding are of various kinds. It is possible to understand subject matters like quantum mechanics, particular states of affairs like the spilling of a cup, institutions like the U.S. House of Representatives, or persons like our best friends.[80] However, Grimm claims that "the differences among these various objects of understanding can be (and have been) overstated, and the reason is that in all these cases understanding seems to arise from a grasp of what we might call dependency relations. Although when it comes to more complex structures (the House of Representatives, for example), *more* of these relations are grasped than when it comes to understanding particular states of affairs; this does not amount to a difference in kind but instead to a difference in degree."[81] Grimm himself does not refer to or cite Baumberger, but instead addresses Kvanvig's account of a distinction of objectual and explanatory understanding and Pritchard's notion of holistic and atomistic understanding.[82] I apply the critique offered by Grimm to Baum-

80 These are the examples proposed by Grimm, see Grimm (2017), p. 214.

81 Ibid. p. 214, original emphasis.

82 Although Pritchard does distinguish holistic and atomistic understanding, he does not provide a detailed account of these forms of understanding. He merely states that atomistic understanding, understanding why such-and-such is the case, is the paradigm usage of 'understanding', that holistic understanding applies to subject matters, and that both usages are related to each other. Pritchard is more interested in the epistemic value of understanding than in its different types. See Pritchard, D. (2010), "Knowledge and Understanding." In Pritchard, D., Millar, A. & Haddock, A., *The Nature and Value of Knowledge: Three Investigations*, pp. 3–90, New York, Oxford University Press, DOI: 10.1093/acprof:oso/9780199586264.001.0001.

berger's notions of explanatory and objectual understanding, since the differentiation in terms of their objects is a crucial characteristic of the two types of understanding in Baumberger' view.

To illustrate his argument that the various objects of understanding only amount to a difference in degree, but not in kind, Grimm presents the two examples of understanding why a cup spilled and understanding the U.S. House of Representatives, which Kvanvig and Baumberger would distinguish as cases of explanatory and objectual understanding. In the first case, Claire sits in a café and observes how the person sitting at the next table accidentally nudges her table with her knee, which causes the shaking of the table as well as of the cup and results in the spilling of the cup. Understanding this event requires Claire to correctly identify the nudging as the cause of the spilling. Claire has to be able to grasp the correct causal relation that led to this event and to omit irrelevant factors such as the time of the day, the music playing in the background etc. In other words, in order to understand the spilling of the cup Claire has to grasp the genuine dependency relation that led to this event (in this case, a causal relation), and not an "empty" or non-causal relation like, for example, between the spilling and the time of the day.[83]

After finishing her coffee, Claire goes back to the library to prepare for an exam. Let us assume Claire is a student in political science and has to learn how the U.S. House of Representatives functions. In contrast to the spilling of the cup, which is a certain event or a particular state of affairs, the House of Representatives is better referred to as a large and complex subject matter. Understanding the House of Representatives means to grasp how the various elements of the institution are dependent on each other, for example "what it takes for bills to be proposed, or for amendments to be introduced, or for them to become laws; who is entitled to speak, at which times; how committees are formed, and how leadership is determined, and so on."[84] Baumberger would argue that this understanding of the House of Representatives is genuinely different than the understanding of the spilling cup, since Claire does not want to understand a particular event via an explanation, but instead a large subject matter via a huge body of information. Grimm disagrees. Just as for understanding the spilling of the cup, to understand the House of Representatives Claire has to be able to grasp the genuine dependency relations between elements of the system and to omit empty relations. Grimm admits that the understanding of

83 See Grimm (2017), pp. 214f. Grimm's idea of grasping the genuine dependency relation between the present factors to understand a certain event instead of „empty" dependency relations could serve as a basis for an account of understanding and misunderstanding. It could be argued that Debbie misunderstood the spilling of the cup if she, for whatever reason, takes the music in the café to be the cause of the spilling. However, the elaboration of this idea will be a task for future work.

84 Ibid. p. 215.

the House of Representatives is more demanding, since many dependency relations have to be grasped and that a visual representation of the grasped dependency relations in the case of the House of Representative would look more like a web, whereas the understanding of the spilling of the cup would look like a singular causal chain. Still, this only amounts to a difference in degree of understanding, not in kind. Even though understanding can vary, on the one hand, with respect to the quantity of grasped dependency relations and, on the other hand, in terms of different foci, namely either on individual nodes (for particular events) or on whole systems, the basis of understanding in all these cases is the grasping of the correct dependency relations. According to Grimm, this fundamental similarity is much more relevant for the concept of understanding than the observed differences, which can easily be accommodated as being the characteristic features for different degrees of understanding.[85]

Khalifa addresses this issue about the object of understanding as well and calls for a *Fair Comparison Requirement.* If one wants to compare objectual and explanatory understanding, the comparison should, for example, take the following form:

a) Lea (objectually) understands the occurrence of the Arab Spring in the early 2010s.
b) Isa (explanatorily) understands how/why the Arab Spring occurred in the early 2010s.[86]

Khalifa accuses proponents of a distinction between the two forms of understanding of frequently making unfair comparisons. For example, they would compare b) with something like

a') Lea (objectually) understands the Arab Spring.

In a'), a different target is addressed as in a) and b). To make a proper comparison to a'), one needs to consider cases like

b') Isa (explanatorily) understands how/why the Arab Spring took place in the way it did.

As soon as fair comparisons are made and objectual apples are no longer compared to explanatory oranges, the distinction between objectual and explanatory understanding seems to disappear.[87]

85 See ibid. pp. 215f.

86 Khalifa also claims that understanding *why* is too narrow to account for explanatory understanding, since answers to some how-questions, like "How does DNA replicate?" are also explanations. See Khalifa (2013), p. 1164.

87 See ibid. p. 1164f.

So, there seems to be little or no hope for a differentiation of objectual and explanatory understanding in terms of their targets. If we pay attention to what precisely the object of understanding is and take understanding to be something like the grasp of correct dependency relations, a view that is not contested by any of the scholars I refer to so far, including Baumberger,[88] a genuine difference of explanatory and objectual understanding in terms of their targets disappears. But what about the other distinction in terms of the vehicles, a theory or explanation, respectively? Can such a distinction be maintained?

### 3.4.3 Can the difference concerning the vehicle be maintained?

This is exactly what Christoph Baumberger & Georg Brun want to argue for, to defend a genuine distinction of objectual and explanatory understanding in terms of the vehicle. They refine their notion of objectual understanding and limit it to the understanding of a subject matter by means of a theory. They explicitly exclude other forms of understanding, like understanding other things than subject matters (e.g. the specific action of a person), understanding via other means than theories, understanding theories themselves and, of course, explanatory understanding, which Baumberger & Brun characterize as understanding why something is the case by means of an explanation. At this point, they address Grimm. Baumberger & Brun argue that, even if there is no genuine difference between the objects of understanding, a state of affairs and a subject matter, in the sense that a subject matter is a very complex state of affairs, the genuine difference in the vehicle of understanding remains.[89] "Objectual and explanatory understanding are also distinguished in terms of the means by which they are achieved. Now, theories enable explanations, but they are not merely sets of systems of explanations. Hence, even if subject matters are simply complex states of affairs, this does not imply that the distinction between objectual and explanatory understanding is spurious."[90] Baumberger & Brun do not adopt a specific account of theories, and view theories to be systems of propositions.

Theories are not themselves explanations, but rather, according to Baumberger & Brun, enable (objectual) understanding and also explanations. Since the authors want to defend a strict distinction between objectual and explanatory understanding in terms of their vehicles, they seem to argue that a subject can gain understanding of a subject matter on the basis of a theory without generating an explanation

88 See Baumberger (2011), p. 79.

89 See Baumberger & Brun (2017), pp. 166f. Note that Baumberger has limited his notion of objectual understanding significantly. In Baumberger (2011), he presented the vehicle of objectual understanding more generally as a comprehensive body of information (a whole theory or account) and the object of understanding as a subject matter, a topic, or a system.

90 Baumberger & Brun (2017), p. 167.

from the theory. Is that a plausible idea of how science produces understanding of the natural world? It is not. As Khalifa argues in his reply to Kvanvig's example of the moving electron, which I address in section 3.3.2, scientists usually employ a theory (e.g. quantum theory) to generate an explanation of the respective phenomenon (the subject matter) they want to understand. Henk de Regt provides a whole account of scientific understanding, according to which understanding is achieved by developing explanations of phenomena on the basis of intelligible scientific theories, and he presents three detailed case studies that make up the basis for his account.[91]

Surprisingly, although Baumberger & Brun claim that objectual understanding through a theory and without an explanation is possible, they themselves present examples of their notion of objectual understanding where the subjects do provide explanations! According to Baumberger & Brun, a scientist who understands climate change must be able to use a climate model in *explanations*, and a philosopher understands issues of medical ethics through an ethical theory if she can provide *explanations* for actual or counterfactual cases.[92] Thus, the notion of objectual understanding from Baumberger & Brun does *not* exclude explanation from understanding! It merely states that an explanation is not the starting point or origin of understanding, in contrast to explanatory understanding. That means, a subject would have explanatory understanding of a subject matter if she receives an explanation as an answer to her question while lacking any theory. And she would have objectual understanding if she uses a theory as a basis for generating explanations. Finnur Dellsén, whose account of objectual understanding without explanation I present in section 2.3, admits as well:

> Although [my] account is not an explanatory account of understanding, it does preserve the kernel of truth in explanatory accounts in so far as a sufficiently accurate and comprehensive dependency model contains the sort of information about a phenomenon that is required to explain it and related phenomena, provided that they can be explained at all. This is so for the simple reason that the dependence relations that these models must correctly represent in order to provide understanding (for example, causal and grounding relations) are precisely the sort of relations that form the basis for correct explanations.[93]

Although Dellsén argues for an account of objectual understanding, he, too, does not deny a strong connection between understanding and explanation. I get back to the issues of Dellsén's account in more detail in chapter six.

---

91 See section 2.1 for a summary of de Regt's account. For more details concerning his account and the case studies, see de Regt (2017).

92 See Baumberger & Brun (2017), pp. 167f.

93 Dellsén (2020), pp. 1282f.

So, scientific understanding is *explanatory understanding* in the sense that explanations are a necessary component of scientific understanding. But this does not mean that scientific understanding is achieved only via explanation. As the examples from Baumberger & Brun and the discussion of Khalifa's view in section 3.3 make clear, taking understanding to be necessarily explanatory does not mean that every component of one's understanding has to be explained or that the understanding is exclusively based on explanations. It only means that *explanations are necessarily somehow involved in the understanding* of a phenomenon. Again, in practice scientists achieve explanation and understanding of a phenomenon via a combination of background knowledge, theories, empirical data and methods. To claim that scientific understanding is, therefore, always objectual is possible *if and only if* the important role of explanation for understanding is appreciated. However, the term 'objectual' understanding is problematic, because it is used to oppose explanatory understanding. As I show in this and the previous section, objectual understanding is often defined or used as a form of understanding that does not require or include an explanation. Both notions of objectual and explanatory understanding, in the sense in which they are usually used in the debate, do not accommodate scientific understanding. Explanatory understanding is too narrow in the sense that it is achieved through an explanation only, objectual understanding gets things wrong if it is construed as having no relation to explanation at all. If taken in these senses, both notions need to be broadened if they are to be applicable to scientific understanding. If both notions are extended, the proposed differentiation between objectual and explanatory understanding becomes insignificant. Hence, a differentiation between objectual and explanatory understanding in the case of scientific understanding of phenomena is superfluous (while it might be useful for other types of understanding).

### 3.4.4 Objectual and explanatory understanding cannot be differentiated

Let me summarize the discussion presented in this section. Baumberger first contrasts explanatory and objectual understanding in terms of their vehicles (explanation vs. theory) and objects (event vs. subject matter). Grimm, in contrast, argues that it is not conclusive to differentiate the two types of understanding in terms of their objects, because structural relations are grasped in both cases. Therefore, the difference between explanatory and objectual understanding is only a matter of degree, but not in kind. Still, Baumberger & Brun object that the difference in the vehicle of understanding, namely an explanation or a theory, remains. Therefore, explanatory and objectual understanding seem to be distinct. However, Baumberger & Brun do claim in the examples they present that subjects need to generate explanations from the theory they use if they want to acquire understanding. Investigations of scientific practice conducted by philosophers of science provide further support for a tight connection of explanation and understanding in science. Additionally, it becomes clear that the prevalent notions of objectual and explanatory understanding, which were meant to exclude or clearly contrast each other, both do not capture scientific understanding. Therefore, a strict distinction between objectual and explanatory understanding is needless for accounting for scientific understanding. Instead, scientific understanding should be conceived of as understanding that necessarily involves explanation, but it is not achieved only or exclusively through an explanation. Theories or comprehensive bodies of knowledge are necessary for scientific understanding as well. Gaining explanations and understanding of empirical phenomena are two interrelated goals of science, and there is no reason to tear them apart.

## 3.5 Why scientific understanding and scientific explanation cannot be torn apart

Is explanation necessary for scientific understanding or is it possible to achieve scientific understanding without any explanation? Proponents of the second option present examples in which understanding is (apparently) achieved without explanation or introduce a conceptual difference between explanatory understanding, which is gained merely through or on the basis of an explanation, and objectual understanding, for which a larger body of information is required. As I show throughout this chapter, none of the proposed examples or accounts of understanding without explanation is convincing in light of a pluralist stance on scientific explanation, which should be adopted as it is more appropriate for accommodating scientific practice.

The crucial mistake that all proponents of a separation of understanding and explanation make is to limit the notion of explanation only to causal or deterministic explanation. This flaw becomes especially obvious in the discussions in sections 3.1, 3.2, and 3.3, where I engage with the positions of Lipton and Kvanvig. Such a narrow view on explanation is neither reasonable nor defendable in light of the vast amount of different types of explanation that can be found in various scientific disciplines, including medicine and biology.[94] Or consider physics, where explanation often invokes conservation laws instead of causes. For example, the moon recedes from the earth because of the conservation of angular momentum.[95] The conservation of angular momentum is a (partial) reason for why the moon is slowly drifting away from the earth, but it is not a cause of the phenomenon. The cause is the difference in the

94 See for example Braillard & Malaterre, (2015), or De Vreese, L., Weber, E. & Van Bouwel, J. (2010), "Explanatory pluralism in the medical sciences: Theory and practice." *Theor Med Bioeth*, 31, pp. 371–390, DOI: 10.1007/s11017-010-9156-7.

95 I thank Martin Carrier for mentioning this example.

rotational speeds of the Earth and the moon. The point I want to make here is that in various disciplinary and historical contexts, in which different and changing norms for acceptable explanations are maintained, various forms of explanation can lead to understanding. Therefore, as also de Regt and Khalifa have argued already, a pragmatic and pluralistic position towards explanation should be taken if scientific understanding, understanding gained in science as a whole, should be accommodated. I do justice to this demand by employing my generic conception of explanation and only require that an explanation must present some reason for some feature of a phenomenon.

In addition, Lipton tries to separate understanding from explanation by arguing that, since understanding is tacit and explanation is always explicit, the two things can be independent from one another. Although I agree that understanding should be conceptualized as tacit in some sense, and I address and explain this topic in detail in the next chapter, I disagree that a *purely* tacit conception of understanding without any relation to an explicit representation is appropriate to characterize scientific understanding of phenomena. This is because scientists want to get things right. They want to discover facts about the world, gain knowledge of the world, construct explanations that capture facts about the world and want to understand the world in the right (true) way. Although we know from the pessimistic induction that scientists can never be sure that they reached the ultimate truth about the world, they want to get as close as possible. Science is an epistemic endeavor. If scientific understanding is completely tacit and unrelated to any explicit representation, may it be explanation or something else, it could not be partially communicated. Consequently, it would not be possible for the individual scientist or her colleagues to assess whether her understanding is appropriate in light of the respective standards of the discipline, as the understanding would not be accessible at all. Scientists want to receive the best possible confirmation that what they understood about a phenomenon is true or that it is justified to understand a phenomenon in this or that way. In order to get the best confirmation possible, understanding should somehow be communicated to or shared with other experts in the respective field, and this is only possible if understanding is not conceptualized as being purely tacit.

Finally, in sections 3.3 and 3.4, I argue that the attempt to distinguish explanatory and objectual understanding in terms of their vehicles and targets does not succeed, either. The shared fundamental assumption is that understanding involves grasping relations of the target that is intended to be understood. Grasping these relations can be more or less demanding, given the different kinds and amounts of relations involved in different targets, given their varying complexity. However, possible differences of the objects of understanding only amount to a difference in the degree, but not in the kind of understanding, as the correct relations have to be grasped in any case. Focusing on the vehicle, the extreme conceptions of explanatory understanding as exclusively stemming from an explanation and objectual understanding as gained via a whole body of information without an explanation both cannot accommodate scientific understanding. Explanatory understanding is too narrow, as scientists (usually) do not understand a phenomenon exclusively through an explanation without any additional information, and objectual understanding does not do justice to the role of explanations in scientific understanding and practice. In order to find explanations of phenomena, scientists need a huge amount of knowledge, ranging from well-established theoretical background knowledge to newly gained empirical data. And, as I already mentioned repeatedly, explanations are pervasive across the sciences and one of their main goals. If a pluralist position concerning scientific explanation is taken, it will be very hard to find examples of scientific research that were conducted and achieved their goals without generating and presenting explanations. Therefore, a discussion on whether scientific understanding is explanatory or objectual in the senses mentioned above is superfluous. Importantly, however, I am making this claim only for scientific understanding of phenomena, leaving open the possibility that other kinds of understanding, such as every day, practical, or aesthetic understanding, may be (more) clearly distinguished into objectual and explanatory forms of understanding. If it is the case that scientific understanding, in contrast to other forms of understanding, always contains both objectual and explanatory components, this characteristic could be the essential feature that makes scientific understanding distinctively valuable or special. However, whether this is or could be the case will be a question for future investigations.

In sum, why does scientific understanding requires scientific explanation? Conceptualizing scientific understanding as requiring scientific explanation can better accommodate both scientific practice as well as fundamental intuitions regarding understanding. Concerning scientific practice, two features are especially striking. First, explanation is omnipresent in science. And second, explaining as well as understanding phenomena are undoubtedly two goals of science, both achieved through conducting research. If scientific understanding is viewed as requiring explanation and explanation is everywhere in science, understanding will be everywhere, too. Hence, when scientists arrive at an explanation, they will also have gained understanding. No matter what comes first, understanding or explanation, there is no good reason to assume that they might be separated in scientific practice. Whether explanation is the result of understanding or the other way around, they will advance together if understanding is tight to explanation. I address my view concerning the relation of understanding and explanation in detail in chapter six. Furthermore, these thoughts are in line with the widely shared intuition of many scholars involved in the debate that usually understanding and explanation are related. If someone understands something, she will be able to explain this thing, or at least some features or aspects of it. Even proponents of a separation of understanding and explanation, like Baumberger and Dellsén, admit that *usually* understanding and explanation are related. If a scientist gains understanding of a phenomenon and cannot make anything of this understanding explicit, the understanding would never be open to scrutiny and would not contribute to science as an epistemic community endeavor. How should such a type of tacit understanding be of value to science? Or in other words, why should we attribute understanding of some phenomenon to a scientist who cannot explain any aspect of the phenomenon? My answer is that we have no reason or justification to attribute understanding to this scientists, and hence should not do so. Turning this into a positive formulation, I argue that scientific understanding should require explanation, because such a conception of understanding makes it externally assessable, and hence valuable for science.

The relation of scientific understanding to scientific explanation is one of the issues on which the accounts of scientific understanding from de Regt, Khalifa, and Dellsén differ. I agree with de Regt and Khalifa in this regard, as the three of us take scientific understanding to be necessarily explanatory. However, the three scholars mentioned also disagree on another crucial and in my view more fundamental topic. What is understanding? De Regt and Dellsén take understanding to be some kind of ability, while Khalifa views understanding to be a type of knowledge. If understanding turns out to be an ability, some kind of know-how that is tacit, like Lipton for example claimed, how exactly does understanding then relate to explanation, which is explicit? These are the topics I address in the next chapter.

# 4. Is scientific understanding an ability?

In the debate on the relation of understanding and explanation that I presented in the previous chapter, an additional concept attracts a lot of attention: abilities (or skills). Peter Lipton refers to abilities in his examples of causation, where subjects are *able* to understand through visual representations or manipulation of systems, and unification, where he refers to Kuhnian exemplars and the scientists' *abilities* to choose and solve new problems that seem similar to the exemplar. According to Jonathan Kvanvig, subjects need to be *able* to grasp relations within or between phenomena. Christoph Baumberger refers to the notion of grasping as well and characterizes grasping as certain *abilities* to perform inferential or counterfactual reasoning, and Steven Grimm specifies grasping as the *ability* to identify the correct relations involved in the object of understanding. Moreover, abilities play a crucial role in the discussion on the nature of understanding. Currently, two options are on the table: either understanding is a type of propositional knowledge, or it is some ability. Among the three accounts of scientific understanding I presented in chapter two, Kareem Khalifa endorses the first option, and Henk de Regt and Finnur Dellsén favor the second. Those who take understanding as being something genuinely different from knowledge carve out this difference in terms of abilities and argue either that understanding is a specific ability, or that understanding at least requires specific abilities that are not necessary for knowledge.

In this chapter, I argue that understanding is an ability. To do so, it is necessary to clarify, first of all, what abilities are and whether they actually are something different from (propositional) knowledge. If it turns out that there is no genuine difference between propositional knowledge and abilities, the discussion whether understanding is an ability or a type of propositional knowledge would be superfluous. Hence, I start with an examination of already existing accounts and analyses of abilities in section 4.1. I address several issues related to the concept of abilities (or knowing-how) and its differences in comparison to (propositional) knowledge (or knowledge-that) in sections 4.1.1 to 4.1.6. On that basis, I develop my own view and definition of abilities in section 4.1.7. I will defend three claims. First, abilities are dispositions to perform some activity successfully with respect to relevant standards. Second, abilities are learned and trained in specific social contexts. Third, the manifestations of abilities are partially tacit, that the manifestation processes can never be fully accessed by the subject who manifested the ability. In section 4.2, I relate my analysis of abilities from section 4.1 to understanding and argue that understanding is an ability to succeed in making sense of a phenomenon, situation, or experience. Therefore, understanding should not be identified with a type of knowledge. Based on Gilbert Ryle and Micheal Polanyi thoughts on understanding, I maintain that understanding, in contrast to propositional knowledge, is gradual, its manifestations are multi-track as well as context-sensitive, and consistent with my definition of ability. Then, in section 4.3, I claim that the process of grasping relations of phenomena and articulating these relations in form of explanations is the manifestation of understanding phenomena. I will conclude in section 4.4 that understanding, while being an ability and, therefore, exceeding propositional knowledge, still requires some knowledge in order to be manifested. Without having some knowledge relevant for the phenomenon in questions, no subject could make any sense of the phenomenon. Understanding and knowledge enhance in conjunction with each other.

In contrast to the previous chapter on the relation of understanding and explanation, the discussion in this chapter is not limited to scientific understanding. Although I develop an account of *scientific* understanding in this book, it is not problematic to discuss understanding in general at this point. Quite the opposite, any characterization of understanding in general can elucidate the nature of scientific understanding in particular, as there will be some commonalities among different types of understanding. What is unique about scientific understanding will be addressed in chapters five and six. But first, let us clarify what understanding is in general. In order to do that, we first need to get clear what abilities are.

## 4.1 What are abilities?

The fact that humans possess various skills or abilities to perform outstanding cognitive or physical performances has always fascinated philosophers. Already Plato and Aristotle differentiated and were engaged with the concept of *technê*, usually translated as skills, craft or art, in contrast to *epistêmê*, which is usually translated as knowledge. And still today, knowing-how or abilities are topics of interest in epistemology and metaphysics. Contemporary debates on skills[1] in analytic philosophy

1 In the last decades, ongoing and flourishing debates on abilities and expertise in various fields of philosophy emerged. Since this dissertation is concerned with the concept of scientific understanding, I cannot consider all aspects and arguments that are of concern in the specialized discussions on skills. For a very recent and extensive overview on the debates on skills and expertise, see Fridland, E. & Pavese, C. (eds.) (2021), *The Routldedge Handbook of Philosophy of Skills and Expertise.* Routledge. For an overview on the concepts of abilities, see

are said to have their origins in Gilbert Ryle's approach to this issue and his basic distinction between knowing-how and knowing-that. Let us start from there.

### 4.1.1 A basic distinction of knowing-how and knowing-that

Ryle's motivation and the main goal of his investigation is to object the then accepted dogma of the 'ghost in the machine'[2], which expects that every valued (i.e. labeled intelligent, clever etc.) practical or material action originates in an internal consideration of regulative propositions. It is only possible to perform an intelligent action if one has thought through the regulative propositions that have an influence on the action. In opposition to this dogma, Ryle argues that intelligent performances are possible without any preceding theorizing and that certain performances, including thinking and theorizing, can in themselves be intelligently exercised. He supports this claim by reference to a vicious regress along two dimensions, which would occur if the differentiation between theorizing and practicing is maintained. First, no intelligent act could ever begin, because considering regulative propositions itself is an act that would have to conform to some regulative proposition that would have to be considered etc. Second, to maintain the strict distinction between theorizing and practical actions creates a gap, which makes it unclear how the intellect might bear on the practice.[3] These two dead-ends of the regress-argument show, according to Ryle, that "to do something [intelligently] (whether internally or externally) is not to do two things, one 'in our heads' and the other perhaps in the outside world; it is to do one thing in a certain manner."[4]

Ryle accuses philosophers of his day of concentrating too much on theories of knowledge that concern the discovery of truths or facts, but which either ignore the methods or ways in which these truths or facts are discovered, or try to reduce the methods to the discovery of facts itself.[5] In contrast, Ryle argues "that knowledge-how cannot be defined in terms of knowledge-that and further, that knowledge-how is a concept logically prior to the concept of knowledge-that."[6] He presents the following example to illustrate his point:

---

e.g. Maier, J., "Abilities", *The Stanford Encyclopedia of Philosophy* (Summer 2021 Edition), Edward N. Zalta (ed.), URL = https://plato.stanford.edu/archives/sum2021/entries/abilities/ (last accessed April 12th, 2022).

2 See Ryle, G. (1949), *The Concept of Mind*. Chicago, The University of Chicago Press, pp. 26ff.

3 See Ryle, G. (1990 [1946]), "Knowing How and Knowing That." In Collected Papers (Volume 2), Bristol, Thoemmes Antiquarian Books Ltd, pp. 212–225, pp. 212f.

4 Ibid. p. 214.

5 See ibid. p. 215.

6 Ibid. p. 215.

> A pupil fails to follow an argument. He understands the premisses and he understands the conclusion. But he fails to see that the conclusion follows from the premises. The teacher thinks him rather dull but tries to help. So he tells him that there is an ulterior proposition which he has not considered, namely, that *if the premisses are true, the conclusion is true.* The pupil understands this and dutifully recites it alongside the premisses, and still fails to see that the conclusion follows from the premisses even when accompanied by the assertion that these premisses entail the conclusion. So a second hypothetical proposition is added to the store; namely, that the conclusion is true if the premisses are true as well as the first hypothetical proposition that if the premisses are true the conclusion is true. And still the pupil fails to see. And so on forever. He accepts rules in theory but this does not *force* him to apply them in practice. He considers reasons, but he fails to reason. [sic][7]

Even in everyday language, the difference between knowledge-how and knowledge-that becomes apparent, so Ryle argues. When we talk about people's beliefs, opinions, or knowledge-that, we ask for reasons or grounds for accepting a certain proposition, but we never talk about someone believing- or opining-how. In the case of knowledge-how, it is different. We can and do describe *how* certain activities are performed, but we do not ask for the grounds or reasons of someone's performance. When we describe *how* people know to, for example, sing or play tennis, we actually mean that they perform those activities well, i.e. that their performances meet certain standards or criteria.[8]

So, Ryle arrives at a fundamental distinction between two kinds of knowledge. On the one hand, there is propositional knowledge or knowledge-that, which covers knowledge of facts, e.g. that light travels with a speed of $3x10^8$ m/s or that Tokyo currently is the capital of Japan. On the other hand, there is knowledge-how, the knowing how to do something. The concept of knowledge-how includes actions or performances, like reasoning, as in Ryle's examples of the pupil who fails to reason while having all the necessary propositional knowledge, calculating, or physical activities like playing a musical instrument, conducting an experiment, or more basic actions like speaking or walking.

How is Ryle's distinction between knowledge-that and knowledge-how related to abilities or skills? Carlotta Pavese observes that "for many tasks at least, it is intuitive that one cannot be skilled at it without knowing how to perform it."[9] I share this intuition. Since Ryle introduced the distinction of two kinds of knowledge, they have

---

7 Ibid. p. 216, original emphasis.

8 Throughout this chapter, I use the terms 'standard', 'criterion' and 'rule' interchangeably.

9 Pavese, C., "Knowledge How", *The Stanford Encyclopedia of Philosophy* (Summer 2021 Edition), Edward N. Zalta (ed.), URL = https://plato.stanford.edu/archives/sum2021/entries/knowledge-how/ (last accessed April 12th, 2022).

become objects of intensive philosophical discussion and furthermore, "the most recent debate on knowledge-how has intertwined with a debate on the nature of skills."[10] As the (potential) difference in kind between knowledge-how and knowledge-that as well as the relation between knowledge-how and abilities remain contested issues until today, a closer look at these concepts is necessary for clarifying what abilities are. In the next sections, I present further details of Ryle's conception of knowledge-how as well as the work from other scholars who engaged with knowledge-how and knowledge-that. More precisely, and in addition to Ryle's analysis, I will refer to the work from Michael Polanyi and Harry Collins, who made important contributions to the issue of knowing-how, in the next three subsections. Sections 4.1.5 and 4.1.6 will then be devoted to a view from virtue epistemology on abilities, and a critique of that view. I will conclude my occupation with the nature of knowing-how or abilities by providing a definition of abilities in section 4.1.7.

### 4.1.2 Knowing-how as unconsciously acting in accordance with rules

In order to arrive at a robust justification for a distinction between knowing-how and knowing-that, a more detailed investigation of knowing-how is in need. Ryle identifies knowing how with knowing a rule. And knowing a rule amounts to the ability to perform an action intelligently, not knowing an extra piece of information in a propositional form. The pupil in Ryle's example mentioned in the previous section knows a lot of logic's rules in their explicit form, but he is not able to argue, i.e. acting in accordance with the rules. An intelligent pupil, by contrast, may have no knowledge of formal logic at all, but might still be good in arguing. The basic problem always remains: a fool can have all the knowledge (-that) without knowing how to act upon these rules, whereas a reasonable person might have never learned any explicit rules but manages to perform in accordance with them anyway. Between knowing rules (knowing-that) and applying or acting in accordance with them (knowing-how) lies a fundamental difference. In Ryle's view, in knowing-how to do certain things, the knowledge of a person is actualized or exercised in what he does, without considering any theoretical propositions. And additionally, the performance of a person is somehow governed by specific rules or criteria that apply to performances of a certain sort, e.g. how to make good jokes.[11] For Ryle, the witty person is able to make good jokes and identifies bad ones, but she will not be able to present any maxims or canons that dictate how she is doing this.[12] Importantly,

10 Ibid. For an overview of classical as well as recent arguments in favor of and against a distinction of knowledge-how and knowledge-that, see ibid.

11 See Ryle (1990 [1946]), pp. 217f.

12 See Ryle (1949), p. 30.

"a skill is not an act. [...] It is a disposition, or a complex of dispositions."[13] And "phrases such as 'technical skill', 'scrupulous conduct' and even 'practical reason' denote capacities to execute not tandem operations but single operations with special procedures."[14]

Intellectualists object to Ryle's position that knowing-how is independent of and prior to knowing that. They argue that, since knowing-how involves knowledge of a rule in every case, the knowing of this rule amounts to the propositional knowledge of a general hypothetical pattern of the form 'whenever so and so, then such and such'. Ryle has two counterarguments against this suggestion. First, knowing and accepting any set of such hypothetical propositions does not automatically imply that they enable a person to successfully act in accordance with them. One might accept the proposition of how to sew a shirt or ride a bike, but this is not sufficient for performing these activities. On the contrary, a person might know how to ride a bike because she practiced it in a trial and error process without ever knowing explicitly what the rules for successful bike-riding look like. And second, the proposed general hypothetical pattern is an inductive generalization. Generating these generalizations requires valid inductive reasoning, which is in itself an intelligent performance that cannot adhere to the general hypothetical propositions that are a result of this performance. We would end up in an infinite regress again. However, Ryle does acknowledge that it is possible to extract propositional principles from successful activities performed by those who know how to perform these activities, like hunting, tailoring or reasoning. These principles are expressed in the imperative form and not in the indicative, though, which implies that we do accept rules or maxims, but, contra to the intellectualists claim, we do not accept any truths behind them, since truths cannot be expressed in an imperative form, so Ryle argues. Still, the extracted principles serve a crucial pedagogical function: they are guidelines or handbooks for novices who are learning certain activities from those who know how to perform these activities successfully.[15]

Persons who know how to do certain things, how to perform activities or give good advice on these activities, are credited with a certain dispositional excellence. This dispositional excellence is actualized in the performed activities. An excellent cook knows how to create delicious dishes without recalling the recipes, a good chess player does not consciously think of the rules and tactical principles of the game, he simply plays according to them, and the acute reasoner does not consider propositions and glances on formula, he simply takes them into account appropriately in his activity of reasoning.[16]

---

13 Ibid. p. 33.

14 Ryle (1990 [1946]), p. 214.

15 See ibid. pp. 220f.

16 See ibid. p. 223.

In Michael Polanyi's view, which sounds similar to Ryle's, performing a skillful action requires "the observance of a set of rules which are not known as such to the person following them."[17] This means that an action can be well or skillfully performed without the consideration of explicit rules. In that sense, skills are unspecifiable. Polanyi explicates this idea by referring to the notions of subsidiary awareness and focal awareness, which he takes to be mutually exclusive. When a pianist is playing a concert, his focal awareness is targeted at the piece of music he is playing at that time. While playing a piece of music on the piano, the pianist is constantly moving his fingers, feet, and his whole body to produce the desired sounds, and these bodily movements lie in his subsidiary awareness. If the pianist would try to focus on what he is actually doing with his fingers, feet and body, he would get irritated and stop playing. The human attention can only hold on focus at a time, according to Polanyi. While performing, the focus lies on the piece that is being played, that is, on the activity as a whole. One cannot simultaneously be focally aware of the whole and of its parts. However, the pianist is subsidiary aware of the parts or particulars, the individual notes and his physical activities to produce these notes, while he is playing the piece. The particulars of a certain activity, like producing every single note on the piano, are unconsciously performed or realized. The whole, the meaning or goal that should be achieved by the activity, playing a piece on the piano, lies in the focal awareness. The coordination and correct application of all the particulars is a demanding and complex process that has to be learned.[18]

> We may say, more generally, that by the effort by which I concentrate on my chosen plane of operation I succeed in absorbing all the elements of the situation of which I might otherwise be aware in themselves, so that I become aware of them now in terms of the operational results achieved through their use. [...] And again, in practical terms, as we learn to handle [certain tools] in terms of the situation which we are striving to master, we become unconscious of the actions by which we achieve this result. This lapse into unconsciousness is accompanied by a newly acquired consciousness of the experiences in question, on the operational plane. It is misleading, therefore, to describe this as the mere result of repetition; it is a structural change achieved by a repeated mental effort aiming at the instrumentalization of certain things and actions in the service of some purpose.[19]

17 Polanyi, M. (1962 [1958]), *Personal Knowledge. Towards a Post-Critical Philosophy*, London, Routledge, p. 51.

18 See ibid. pp. 57–68.

19 Ibid. p. 64.

Harry Collins provides another, slightly different analysis of 'tacit knowledge', as he calls it.[20] Although Collins uses a different terminology, the notions 'tacit knowledge' and 'knowledge-how' refer to the same kind of knowledge, namely knowledge that is not or cannot be articulated explicitly. He distinguishes three different kinds of knowledge that are often labelled 'tacit': relational, somatic, and collective tacit knowledge. While the first two can be made explicit in principle, the latter remains inherently tacit.[21] The crucial dimension that makes collective tacit knowledge distinct from the other two kinds, and also from explicit knowledge, is the social dimension, according to Collins. He clarifies this idea by discussing the two activities of bicycle riding and bicycle balancing.[22] Bicycle balancing requires the coordination of neural and muscular acts to stay upright on a bike and to move forward. Bicycle balancing is an instance of somatic tacit knowledge, as it can be made explicit through scientific research. The additional component involved in bicycle riding is the social component since activities like bicycle riding take place in a social environment like traffic. In any social setting, not only in traffic, the subject has to make a social judgement about the balancing of the individual and social responsibility in order to keep the setting working.[23] To put it differently, Collins argues that knowledge is ultimately located in society, and individuals merely share the knowledge of the collective in an implicit, tacit way. How humans manage to participate in collective knowledge and connect to society exactly is still obscure. This is the feature that makes collective tacit knowledge inherently tacit.[24] Collins' view in a nutshell, to obtain a certain skill requires the respective collective tacit knowledge. Collective tacit knowledge is a prerequisite for having skills. First a person needs to gain collective tacit knowledge, and then she can train a certain ability that accords with it. Without

20 See Collins, H. (2010), *Tacit and Explicit Knowledge*. Chicago and London, The University of Chicago Press.

21 Relational tacit knowledge can be made explicit in principle and only logistical or social reasons hinder persons to do so. Collins discusses several instances of relational tacit knowledge, e.g. cases of mismatched saliences in which some fact is not made explicit because person A takes for granted that person B knows this fact, while she actually does not. Somatic tacit knowledge is tacit for the individual because of the limits of the human body and brain, but it can be explicated in terms of scientific theories or explanations. Therefore, somatic tacit knowledge, as well as relational tacit knowledge, is not an instance of entirely tacit knowledge because it can be made explicit in principle by scientists. For more details concerning relational and somatic tacit knowledge, see ibid. pp. 85–117.

22 Collins discusses bicycle balancing and chess playing as examples for somatic tacit knowledge. The riders of a bicycle or champions in chess are not able to state explicitly and in every detail how they are riding a bicycle or played a certain chess game at a tournament, although they are able to apply that knowledge in a practical sense. However, science can make this knowledge explicit, see ibid. pp. 99–117.

23 See ibid. pp. 119–123.

24 See ibid. pp. 131–138.

the acquisition of some collective tacit knowledge, subjects could not acquire abilities. Collins' distinction between collective tacit knowledge and abilities resembles Ryle's and Polanyi's analysis, as they also claim that subjects need to possess tacit or implicit rules, that they have the knowledge-how, to manifest skillful performances.

So, according to the three authors, knowing-that is possessing explicit information or facts, while knowing-how is a disposition or capacity to act in accordance with certain tacit rules. Once a subject acquired some tacit rules, she will perform the respective activity well without consciously or explicitly considering any rules. In contrast to knowing-how, knowing-that consist of the possession of explicit facts. If the two types of knowledge differ in the way just described, would the process of learning or acquiring them also vary significantly? Some argue that it does. If this is the case, a differentiation between knowledge-how and knowledge-that as two distinct kinds will become even more plausible.

### 4.1.3 The acquirement of knowing-how

Unsurprisingly, Ryle is among those scholars who argue that there is a fundamental difference between how we acquire knowledge-how and knowledge-that. According to him, and I am paraphrasing Ryle's way of talking here, the latter involves the instruction of truths and the accumulation of pieces of knowledge-that, of facts. In contrast, knowing-how requires appropriate exercising and being disciplined in methods. The term 'discipline' refers to two different processes, habituation or drill on the one hand, and education or training on the other. Ryle is not concerned with habituation or drill. This measure produces automatisms and blind habits. In drilling novices, they are learning to do something blindly without considering how they are performing their task or why they are performing their task exactly like this or what alternative realizations of the task could be. Contrary to that, Ryle claims that educating or training novices, the second meaning of 'discipline', enables them to develop intelligent powers. In teaching skills, novices learn how to do something intelligently. Training supports and fosters intelligence, whereas drills dispense it. The education of novices allows them to perform exercises in the right way by using their brains, to learn from their mistakes and how to avoid or correct them. Although skills include habits, skills exceed habits in that they allow for the generation of new successful procedures as well as of new propositional knowledge, instead of merely repeating the same activity time and again. Disciplines such as mathematics or philosophy or methods from the sciences could be seen as branches of knowledge-how, not (just) bodies of propositional information.[25] "The experts in them cannot tell us what they know, they can only show what they know by operating with cleverness, skill, elegance or taste. The advance of knowledge does

25 See Ryle (1990 [1946]) pp. 223f.

not consist only in the accumulation of discovered truths, but also and chiefly in the cumulative mastery of methods."[26] Accordingly, "we learn *how* by practice, schooled indeed by criticism and example, but often quite unaided by any lessons in the theory."[27]

Polanyi agrees with Ryle on the nature of knowledge-how and on its acquisition process. For Polanyi, science is a form of art and the art can be passed on only by example from master to novice. This is so because the rules that have to be subconsciously observed to perform skillfully are not explicitly known by any subject, cannot define the respective art in detail, and cannot be conveyed in the form of instructions. Personal contact is required to learn an art. In the case of science, the explicit contents, like theories or explanations, were distributed globally much faster than the art of conducting scientific research.[28] "The large amount of time spent by students of chemistry, biology and medicine in their practical courses shows how greatly these sciences rely on the transmission of skills and connoisseurship from master to apprentice. It offers an impressive demonstration of the extent to which the art of knowing has remained unspecifiable at the very heart of science."[29]

Finally, the necessity of personal contact in the process of acquiring knowledge-how or tacit knowledge, is also stressed by Harry Collins. As it is the social dimension of collective tacit knowledge that makes it inherently tacit, collective tacit knowledge cannot be acquired if one is not situated in a society. How to do things right in a specific social setting cannot be learned by rules, according to Collins, but only by experience. Subjects need to absorb social rules that cannot be made explicit, and they can do so only by being part of a society.[30] Collins provides the example of dance improvisation to illustrate this point. "Improvisation is a skill requiring the kind of tacit knowledge that can *only* be acquired through social embedding in society. Social sensibility is needed to know that one innovative dance step counts as an improvisation while another counts as foolish, dangerous, or ugly, and the difference may be a matter of changing fashion, your dancing partner, and location."[31] Tacit knowledge, knowledge-how, can be acquired through language, through the derived pedagogical principles, but also through physical activity, whereupon Collins takes the participation in physical activity to be the more efficient way for doing so.[32] This is

26 Ibid. p. 223.

27 Ryle (1949), p. 41.

28 See Polanyi (1962 [1958]), pp. 51–56.

29 Ibid. p. 57.

30 See Collins (2010), pp. 119–123.

31 Collins (2010), p. 123, original emphasis.

32 See ibid. pp. 131–138. For more details concerning the advantage of physical activity over mere conversation see, for example, Ribeiro, R. & Collins, H. (2007), "The Bread- Making Machine, Tacit Knowledge, and Two Types of Action." *Organization Studies*, 28 (9), pp. 1417– 1433, DOI: 10.1177/0170840607082228.

consistent with Ryle's claim that propositional knowledge alone is not sufficient for performing any corresponding skillful action and with Polanyi's point that learning skills requires interaction with another, skillful person.

Summing up, how is knowing-how acquired by subjects? How do we learn to do something correctly, or intelligently, or well in any sense? According to Ryle, Polanyi, and Collins, acquiring knowing-how requires learning the tacit social rules and practicing performances that accord with these rules in a social setting. We need to acquire methods and train them by exercising them. Acquiring knowledge-how is a learning by doing procedure that is guided by a master or a social community, who gives the subject feedback on her progress with the method or practice in question. Assessing whether the tacit rules that govern practices have been incorporated by the subject requires other members of the community. Although the novice receives feedback during her training, nobody can tell her explicitly what she needs to do. Ultimately, she has to acquire the knowledge-how herself, through her own experience. This process of acquiring knowledge-how is fundamentally different from acquiring knowledge-that, as in the second case, explicit information or facts are gathered and structured. Knowledge-that requires the possession of explicit information, knowledge-how the awareness of tacit rules. Therefore, not only the nature of the two types of knowledge is fundamentally different, but also their acquisition processes. This is the conclusion at which Ryle, Polanyi and Collins arrive concerning the acquisition process of knowing-how, and I consent to this view.

### 4.1.4 Manifesting knowing-how is context-sensitive

While having the know-how to perform an activity well requires the possession of tacit rules, it does not amount to simply repeating certain actions in every performance. I already mentioned that Ryle distinguishes between automatism and blind habits on the one hand, and intelligent powers or capacities on the other hand. As an illustration of how intelligent capacities allow a person to modify his performance, Ryle discusses what happens when a person is arguing intelligently. The intelligent reasoner does not simply repeat his argument again and again, as it would be the case if his ability to argue intelligently would merely be a habit, but instead constructs new parts of his argument that did not exist before, i.e. that he did not consider before. This construction of a new argument is due to changing requirements, which occur in every new case of arguing: the meeting of new objections, interpreting new evidence, making connections between elements in the specific situation which did not correspond before. While the intelligent reasoner constructs new arguments and learns with every new case of argumentation, he always reasons logically. The rules of logic are observed in such a way that the intelligent reasoner does not consciously consider them. Instead, the rules of logic became his way of thinking. The intelligent reasoner reasons in accordance with a specific method, but without reflecting on the prescriptions of the corresponding methodology. The same holds for physical practices, e.g. being a good surgeon. Without any medical knowledge, a person will never become a good surgeon but having a lot of explicit medical knowledge itself is not sufficient for being a good surgeon. A surgeon must have learned the practice of performing surgery, and that might require slightly different decisions on actions and steps during performing surgery on different patients.[33]

Polanyi and Collins can be aligned with Ryle's view on the openness of skills. Polanyi takes the rules of art, which a subjects follows during a performance, as useful and as a guide to an art, while not dictating the practice of the art.[34] Collins' conception of collective tacit knowledge fits well into this picture, too. In order to achieve the respective collective tacit knowledge, a subject needs to have a social sensibility. This, in turn, requires a brain that can assimilate social rules. Since social rules might change with context, the process of learning skills must be flexible. This is different to the training of muscles or habituation, as Ryle calls it. Balancing a bike works the same everywhere, but how to ride a bike in traffic differs with respect to society, that is, to different traffic systems. If a subject wants to learn a skill, she will have to acquire the respective tacit knowledge and to do that, she needs to stay in touch with society. Collins concludes that skill acquisition occurs in two steps or phases. First, the mere acquisition of motor coordination, and second, the acquisition of motor coordination in a socially sensitive way.[35] The collective nature of tacit knowledge and hence of skills is irreducible, because "it is only humans who have the ability to acquire cultural fluency. It is only humans who possess what we can call "socialness" – the ability to absorb ways of going on from the surrounding society without being able to articulate the rules in detail."[36]

Hence, another crucial difference between knowing-how and knowing-that is revealed. I conclude that the manifestation of knowing-how is context-sensitive, it can be changed or adjusted if required. Showing some knowing-how is not mere repetition or habit. Importantly, no matter what exactly a subject changes, the modified performances will still be in accordance with the tacit social rules if performed successfully. Independent of the concrete execution, the skillful performance will always count as a skillful performance. Knowledge-that is not context-sensitive in this way. We take knowledge-that, like the facts that light travels with a speed of $3x10^8$

33 See Ryle (1949), pp. 46–49.

34 See Polanyi (1962 [1958]), pp. 51f.

35 See Collins (2010), pp. 123f.

36 Ibid. p. 125. Importantly, Collins does not claim that animals are not social. All he is saying is that, for example, animals eat their foot in the same manner all around the world, while humans can change their eating manners according to the cultural setting. Even if Collins is wrong in this regard, attributing "socialness" in this sense to humans as well as animals will not affect the ability of humans to modify their performance according to the context.

m/s or that Tokyo currently is the capital of Japan, to be true independent of any context. There is no room for the same kind of variation or interpretation in the case of knowledge-that that as for the manifestation of knowledge-how.

### 4.1.5 Knowing-how, abilities or skills are stable, robust features

Although the manifestations of abilities are context-sensitive and exceed repetition or habit, as skillful persons can modify their performance if necessary, they also seem to have a stable or robust component. Able dancers, bicycle riders or chess players are called 'able' because they were and are very often successful in their performances. This idea is elaborated by two virtue epistemologists, namely Ernest Sosa and John Greco. Both conceive knowledge as an intellectual success achieved through ability,[37] which again is compatible with the view presented before that knowing how precedes propositional knowledge. Sosa argues that competences, which he identifies with dispositions of an agent to perform well, include three components: its constitution, its condition, and its situation. It is the constitutional competence that Sosa identifies as a skill. A ski jumper, for example, has learned all the skills he needs to be a good ski jumper during his athletic career, he has internalized them. However, to execute a good jump, i.e. to be successful in what he is doing, he needs to be in the right condition, well-trained and prepared for the respective competition, healthy, sober etc. And finally, he needs the favorable situation, good weather conditions, to perform well. The best ski jumper will not be able to execute a good jump in a thunderstorm. The same structure holds for intellectual competences. A theoretical chemist must learn all the skills required for the discipline during her studies. In order to make accurate calculations to predict chemical phenomena, she also needs to be in the right condition (well-rested, focused, sober) and in the right situation. Even for the best theoretical chemist it might be impossible to perform sophisticated computations in an office with 35°C with no air conditioning. The crucial aspect of a skill, the constitutional part of a competence, is, according to Sosa, that the agent always retains the skill, even when she is asleep, ill, or drunk. That is, an agent is always in possession of certain competences even if she is not able to perform these competences well in all conditions or situations.[38] "That you fail a conditionals test when in poor shape or

---

37 For an overview on virtue epistemology, see for example Turri, J., Alfano, M. & Greco, J., "Virtue Epistemology", *The Stanford Encyclopedia of Philosophy* (Winter 2021 Edition), Edward N. Zalta (ed.), URL = https://plato.stanford.edu/archives/win2021/entries/epistemology-virtue/ (last accessed April 12th 2022).

38 See Sosa, E. (2010), "How competence matters in epistemology." *Philosophical Perspectives*, 24, pp. 465–475., DOI: 10.1111/j.1520-8583.2010.00200.x, p. 465.

poorly situated bears not at all on your possession of a corresponding constitutional competence."[39]

John Greco, who studied under Ernest Sosa, also identifies intellectual virtues with abilities or powers of the knowing agent. In his view, knowledge is true belief based on intellectual ability. Starting from this, Greco can explain why knowledge is *creditable* true belief since knowledge is a form of success through ability, and success through ability is worthy of a special sort of credit.[40]

> More generally, the sort of crediting and valuing associated with success through ability (or excellence, or virtue) is ubiquitous in human life. It is instanced in the moral realm, the athletic, the artistic, and many more. [...] We credit success through ability more than we credit mere lucky success. But we also *value* success through ability more than we value mere lucky success, i.e., success through virtue or excellence, is identified as the highest human good: it is of intrinsic value itself, and it is constitutive of human flourishing.[41]

If an ability is attributed to an agent, it will mean that she is reliably successful in some way relevant to the respective ability, according to Greco. Moreover, abilities are tethered to significant conditions and the environment. Greco's distinction between condition and environment is not identical to Sosa's differentiation between condition and situation, since Greco takes the concepts of "condition" and "environment" to be overlapping. The only vague distinction that Greco offers is that "environments" refer to sets of relatively stable circumstances, while "conditions" indicate sets of shifting circumstances within an environment.[42] "Finally, to say that someone has an ability to achieve some result is to say both more and less than that they have a good track record with respect to achieving that result. This is because abilities are *dispositional properties*: to say that S has the ability to achieve result R is to say that S has a disposition or tendency to achieve R across relevantly close worlds."[43] That is, abilities as well as their attribution are context-sensitive. Which conditions and environments, that is, which close worlds, are seen as relevant depends on the interests and purposes that are operative in the respective context. If,

---

39 Ibid. p. 469.

40 See Greco, J. (2007), "The Nature of Ability and the Purpose of Knowledge." *Philosophical Issues*, 17, pp. 57–69, DOI: 10.1111/j.1533-6077.2007.00122.x, p. 57.

41 Ibid. pp. 57f, original emphasis. Greco develops his account with the goal of explaining why knowledge is incompatible with luck and why knowledge is more valuable than mere true belief. For a more advanced presentation of his account, see Greco, J. (2010), *Achieving Knowledge. A virtue-theoretic account of epistemic normativity*, Cambridge, Cambridge University Press, DOI: 10.1017/CBO9780511844645.

42 See ibid. p. 60.

43 Ibid. pp. 60f, my emphasis.

for example, Oliver states that Thomas has the ability to coach a soccer team well, this statement may have a different meaning, depending on whether Oliver refers to Thomas Tuchel, the coach of the soccer club Paris Saint-Germain, or Thomas, the coach of Oliver's ten years-old son. The ability needed to build a functioning team out of individual soccer geniuses playing on the highest international level is quite different from the ability to coach ten years-old boys in a local league. In these two cases, the relevant conditions and environments are fundamentally different. Greco allows for subject, attributor, or third party contextualism, depending on whose interests and purposes are of concern.[44]

So, once a subject acquired an ability, some knowing-how, this ability will be a constitutional or at least a stable feature of the subject, which means that she has been and will be reliably successful in manifesting this ability in relevant and appropriate contexts. This insight is in line with the result of the previous section: while the ability is a stable, robust feature of a subject, the manifestation of this ability will be sensitive to the given context and, hence, might differ, respectively.

### 4.1.6 Should knowing-how be tight to success?

Barbara Vetter summarizes Sosa's and Greco's characterization of abilities or competences as "dispositions to succeed".[45] This idea can already be traced back to Ryle, who conceived knowing-how to perform actions as performing these actions well or successfully with respect to certain criteria. Polanyi and Collins approve of this, as I showed. The crucial point about this conceptualization of abilities is that agents will perform a certain action properly if they do it at all. That is, a ski jumper would perform a good jump if he jumps, given an appropriate condition and situation, and Thomas would be a good (able) soccer coach if he were to coach a soccer team.[46] Vetter characterizes Sosa's and Greco's account of abilities as follows:

> *x* has the ability to A iff *x* is disposed to A successfully when A'ing at all, i.e. iff, if *x* were to A at all, then (interferences aside) *x* would (probably) A successfully.[47]

However, Vetter identifies and discusses a crucial problem of this account of abilities. According to Sosa and Greco, the agent's exercise of the ability is tied to success. But what about activities that do not have standards of success, that cannot be performed 'well' in any sense? Aimless ambling and doodling are examples of perfor-

44 See ibid. pp. 60ff.

45 Vetter, B. (2019), "Are abilities dispositions?" *Synthese*, 196, pp. 201–220, DOI: 10.1007/s11229-016-1152-7, p. 214.

46 See ibid. pp. 214f.

47 Ibid. p. 214.

mances which lack any aim as well as any evaluative standard. One cannot amble or doodle in a better or worse way. Either one does amble or doodle, or one does not. It is possible, though, to introduce a trivial success condition: The end of aimless doodling is aimless doodling. If my random doodling ends up in a sketch of my boss, I will have failed to doodle randomly. If my aimless ambling turns into a walk to the train station because I spontaneously want to visit a friend, I will have failed to amble aimlessly. Furthermore, Vetter notes that many verbs, like 'hit', 'move' or 'reach', describe a performance only in so far as it is performed successfully. Therefore, they are called 'success verbs'. Vetter concludes that Sosa's and Greco's characterization of abilities as dispositions to succeed becomes trivial in the cases of success verbs, of aimless activities, and also of simple motor activities (like raising one's arm, moving one's eyes, or wiggle one's foot), because in these cases, A'ing automatically amounts to A'ing successfully.[48]

The consequence of this trivial characterization of dispositions as being disposed to A if A, or that one would A if one were to A, is that either everything has these dispositions, and hence the respective ability (everything can hit a board, amble aimlessly or move something), or that nothing has such dispositions, and the corresponding ability, if there are no such trivial dispositions. Both alternatives are absurd, according to Vetter.[49] Either one would have to accept that, in the case in which a sponge hits the blackboard, the sponge has the disposition to hit the blackboard, and, therefore, the sponge has the ability to hit the blackboard. No one would attribute abilities to sponges, at least not to the ones we find in old-school classrooms to clean the blackboard! Or we would have to deny that dispositions like hitting a blackboard exist. If we do that and if, at the same time, we side with Sosa'a and Greco's characterization of abilities as dispositions to succeed when you try, we could rule out absurd attributions, but we would have to deny that anything or anyone has the disposition, and therefore the ability, to hit the blackboard. Ergo, no one would be able to hit a blackboard, which is definitely not true, as some students might frequently hit the blackboard when throwing a sponge at it, while others might often fail to hit the blackboard with the sponge. Given these problems with 'abilities as dispositions to succeed', Vetter concludes that this "account failed to ensure that the manifestation for the correlated disposition is always sufficiently distinct from the performance itself and thereby fails to capture at least a great deal of our simple motor abilities."[50]

48 See ibid. pp. 215f.

49 See ibid. p. 215.

50 Ibid. p. 217. In general, the notion of manifestation refers to the process or event of becoming visible or the revelation of all kinds of things, which were invisible or shapeless or even non-existent before the manifestation took place. For an overview of topics concerning dispositions and manifestations that are addressed in philosophy, see for example Choi, S. & Fara, M., "Dispositions", *The Stanford Encyclopedia of Philosophy* (Spring 2021 Edition), Edward

Generally, Vetter is interested in the nature of abilities and starts her own investigations with the insight that abilities are commonly referred to as dispositional properties, an idea that can also already be found in Ryle's work. However, not every disposition would ordinarily be considered an ability, like the dispositions to get provoked or get a sunburn easily. And the question which dispositions are abilities is still not answered conclusively. This is the context in which Vetter discusses the views of virtue reliabilists like Sosa and Greco. She wants to show that this view fails to fully capture the intuitive notion of ability, but she does not provide an alternative answer to the question which dispositions are abilities. It might even be possible that there is no unified, reasonable conception of ability that captures all sorts of abilities. Rather, it could be that the term 'ability' covers are large number of (partially) overlapping meanings. However, if this is the case, these meanings need to be worked out in detail, which has not been done yet. To conclude, on the one hand, it is still not clear what it means to have an ability. Yet, it is obvious that agents have abilities. Agents are able to vote, drive a car, and do all kinds of things.[51] As Vetter puts it, "our abilities – unlike the opportunities with which the world presents us – tend to be stable, robust features of ourselves that we can rely on in a large variety of different situations. The ability [...] is there to be called upon, even when it is presently lying dormant, that is, unexercised."[52]

### 4.1.7 Abilities are dispositions to succeed

Where does this discussion of abilities leave us, what are abilities? I am now in a position to present my own view on that matter. Although Vetter's critique is a challenge to any general account that characterizes abilities as dispositions to succeed, it is not a problem for activities that do have standards of success and which do serve an aim. Since she does not propose any alternative conception of ability and does allow for the possibility of several and potentially overlapping conceptions of abilities, I do not see a problem in sticking with the idea that some abilities are disposition to succeed, or to perform well or in accordance with some given standards. Again,

---

N. Zalta (ed.), URL = https://plato.stanford.edu/archives/spr2021/entries/dispositions/ (last accessed April 12th, 2022).

51 See Vetter (2019), pp. 201f, 218f. In addition to the view of virtue reliabilism on abilities, Vetter also discusses an alternative approach. Namely, the project called 'new dispositionalism' from the debate on free will, which states that abilities are dispositions to do what one intends to do. Discussing the arguments from the new dispositionalism here would lead to far and Vetter identifies severe problems with this view of abilities, which I take to be more serious than her criticism of virtue reliabilism. In short, new dispositionalism cannot account for abilities that are incompatible with trying, like creative or sub-intentional actions. For more details, see ibid. pp. 205–214.

52 Ibid. p. 202.

as Vetter notes, this characterization does not accommodate all abilities that agents possess, but it does accommodate those kinds of abilities I am interested in in the context of scientific research. I agree with Vetter that the questions what abilities are and whether one unified conception of ability is possible are interesting and important. However, they lie outside of the scope of this book.

Let me quickly turn to the inconsistent terminology used by the various authors I refer to. Throughout this section, terms like knowing-how, skill, ability, or competence appear. Ultimately, I think that the terminological variation does not point to a fundamental metaphysical difference. The authors I address in this section employ expressions like 'someone has the ability to x, has the skill to x, has the know-how to x, or has the competence to x' to refer to the same thing. They all denote that someone can do something in an appreciated or valued manner. The only concept that slightly steps out of line is 'disposition', since we would not call every disposition an ability. Just because a sugar cube has the disposition to dissolve in water, no one would say that sugar cubes are able to dissolve in water. Hence, not all dispositions are abilities. But if we focus on those dispositions that have standards of success, we can say that the terms 'ability', 'skill', 'know-how', and 'competence' can all be defined as denoting someone to have a disposition to perform some activity successfully with respect to relevant standards.[53]

Fortunately for my purpose, science is a context that has established standards of success, albeit different ones in different disciplines, which have to be met when performing certain activities in the course of scientific research. A study or experiment can be set up and conducted well or badly, a specimen can be prepared well or badly, a laser system can be adjusted well or badly, a questionnaire can be devised well or badly. The kinds of abilities that are not captured by the 'abilities as dispositions to succeed'-view, performances described by success verbs, aimless activities, and simple motor activities, are not the kinds of abilities that play an important or central role in scientific research. All abilities or skills that scientists learn over the course of their training are highly specialized abilities, which ultimately serve the aims of conducting good science, discovering truths about the world, understanding and explaining natural and social phenomena, making correct predictions, improving people's lives and possibly even more or different aims. If the highly specialized skills that scientists acquire would not serve some purpose, scientists *qua* scientists would not learn and train them in the first place. Hence, the 'abilities as dispositions to succeed'-view suits the characterization of the very specific and purposeful skills learned and exercised by scientists. Moreover, the 'abilities as dispositions to succeed'-view also accommodates many activities outside of the scientific realm.

---

53 If it turns out that I wrongly equate these terms, I will happily only use 'ability' and 'disposition (to succeed)' and refrain from using 'skill' and 'competence'. However, I am not aware of a good reason or argument to do so.

Throughout this section, I mention examples from various domains, including being an able, and hence reliably successful, cook, chess player, bicycle rider, surgeon, reasoner, or athlete. As these performances are not characteristic for science, except for reasoning, they will have to meet standards of success different from those found in science. That different performances must meet very different standards of success in different domains or realms, however, is not a problem for the 'abilities as dispositions to succeed'-view, as this view does not prescribe any specific standards that must be met. It only implies that some standards of the specific context in question have to be met. Therefore, while science certainly is one prominent example of a domain with established standards for determining successful performances, it definitely is not the only one.

So, the abilities that scientists acquire in their training are dispositions to succeed in what scientists do when exercising those abilities. Furthermore, as Sosa and Vetter emphasize, abilities are stable, robust, and constitutional features that, once acquired, are permanently possessed by the subject even when they are not manifested. In order to acquire an ability, that ability must be learned and trained. And to learn an ability is to learn the respective tacit, non-propositional rules that guide certain actions. As Ryle, Polanyi and Collins argue, the relevant rules can only be learned by participating and practicing in a social setting and not from a rulebook, due to their tacit nature. For science, this amounts to young researchers learning from the example and criticism of their supervisors and more experienced colleagues by actively participating in the scientific community. Although there is no guaranty that scientists with certain abilities will always be successful, since certain conditions or situations might prevent them from succeeding, they will be successful in the appropriate or sufficiently standardized context. Yet, according to Ryle, it is an important feature of skillful performances, of the manifestation of abilities, that they can be changed or adjusted in cases where the context requires changes or adjustments, while continuing to act in accordance with the tacit rules that prescribe how performances in scientific research are to be done. And recall, possessing an ability means to act in accordance with the rules without actively reflecting on these rules. Ryle calls this the internalization of the rules, and Polanyi accommodates this aspect with his discussion of subsidiary and focal awareness. When a scientist is conducting an experiment or performing a calculation, her focal awareness is on the experiment or on the calculation. Without being consciously aware of it, the able scientist performs skillful research in accordance with the scientific rules of her discipline.[54]

54 The tacit nature of abilities might actually be a problem for the new dispositionalism, the second view that Vetter discusses in addition to virtue reliabilism and which I do not address. But I want to mention here that if, according to the new dispositionalism, an ability is a disposition to do X if one intends (or tries, or chooses, or decides) to do X, then the exercise of an ability must always be preceded by an intention, attempt, choice, or decision. Vetter, who

Based on the discussion of abilities in this section, I propose and use the following definition of ability:

x is an ability if and only if x

i. is a disposition to perform a cognitive or physical activity successfully with respect to relevant standards,
ii. has been learned and trained in a specific social context, and
iii. manifests in processes that are partially tacit (i.e. that can never be made fully explicit).

In saying that the manifestations of the abilities are partially tacit, I mean that no subject will be able to figure out and describe precisely how she managed to manifest an ability. This claim is not only in line with Ryle's idea that the rules of a specific performance become the way of acting of the skillful person without any conscious reflection, but also with Polanyi's differentiation between subsidiary and focal awareness and with Collins' notion of collective tacit knowledge. Consider the two examples of dance improvisation and logical reasoning. The former is an instance of the manifestation of a physical ability, the latter of a cognitive ability, an ability that takes place purely in the mind of a subject. In both cases, no subject will be able to accurately describe in all detail which muscles or which neurons did what at which point in time during the performance. No one will be able to describe in detail how he managed to come up with some form of improvisation that he never knew of before, or how she was able to construct a syllogism in her mind that she never considered before. While we are focused on demonstrating a mesmerizing innovative dance performance or on constructing a valid and sound argument, we are focused on performing. But we are not and *cannot* be aware of how we manage to coordinate all the activities of our body and mind, which are necessary to perform well, in accordance with the social rules at play. Note, it is not tacit for us whether we did carry out a specific performance in the end, whether our performance was skillful

does not directly refer to a tacit nature of abilities, provides convincing examples that, first, agents can exercise plenty of abilities without trying, intending, choosing or deciding to do so in advance. And second, that some abilities are only exercised when the agent does not try, choose, decide, or attempt to exercise them, like creative abilities. An agent might have the intention to present a dance performance fantastically, to play the piano with virtuosity, or to conduct an experiment accurately, but while she is focusing on this 'bigger' performance, she cannot pay attention and cannot intend to perform any tiny movement, adjustment, or reasoning step that is involved in presenting a dance performance, playing the piano with virtuosity, or conducting an experiment accurately.

or not. Imagine a cheering crowd around the dance floor and the judges award you a high score for your performance, or you presented a logical argument that your supervisors and peers accepted as a correct and maybe even a good argument. Then we can be pretty confident that we performed the activity in question successfully with respect to the relevant standards that we learned by being in the respective social context that is decisive for the performance in question. As these social standards or rules are implicit themselves, we can neither explicate the rules themselves, nor how we managed to act or perform in accordance with them.

Importantly, I am not claiming that abilities are always independent from knowledge-that or that the two are mutually exclusive! In order to acquire or manifest some ability, subjects may actually need some propositional knowledge. Propositional or explicit knowledge may be necessary for some abilities. For example, a first semester student in philosophy may first have to gain explicit knowledge about what a syllogism is and how to construct one, before she can start to practice and train constructing syllogisms herself. However, I argue that having some propositional knowledge is neither sufficient for also having some ability, nor for manifesting that ability. The philosophy student might have the (propositional) knowledge what a syllogism is and how it is constructed, but when asked to demonstrate this and construct a syllogism herself, she may be unable to do it. In case she acquired the ability to construct syllogisms, she will be able to do this, to manifest the ability, without consciously reflecting on the propositional knowledge of how this is to be done. She will simply do it without explicitly considering how she is doing it. This is the sense in which I take knowledge-that and knowledge-how to be distinct kinds.

After arguing that knowing-how is fundamentally different from propositional knowledge and having arrived at a definition of abilities, I am now in a position to address the central question of this chapter. Is understanding an ability, a kind of knowing-how, or is it more plausible to conceptualize understanding as a type of propositional knowledge?

## 4.2 Understanding, an ability in itself

In chapter two I indicated that there are two opposing camps concerning the nature of understanding. One camp, including Kareem Khalifa and Peter Lipton, takes understanding to be a kind of (propositional) knowledge, the other, with such members as Henk de Regt, Jonathan Kvanvig or Christoph Baumberger, views understanding as a specific ability. In this section I will side with the latter and argue that understanding should be conceptualized as an ability: It is the ability to make sense of experiences, situations, or phenomena in the world. Importantly, the analysis in this section concerns understanding in general and is not limited to scientific understanding. However, if I succeed in showing that understanding is reasonably captured by my definition of ability, specific types of understanding, including scientific understanding, will likely be accommodated by my definition as well. To get started, let us take another look at the authors I was concerned with in the previous section.

### 4.2.1 Early views on understanding as an ability

Already Gilbert Ryle was concerned with the notion of understanding in his investigation on knowledge-how and asked: "What is this difference between merely witnessing a performance and understanding what is witnessed? What, to take another example, is the difference between hearing what a speaker says and making sense of what he is heard to say?"[55] According to Ryle, "understanding is a part of knowing *how*. The knowledge that is required for understanding intelligent performances of a specific kind is some degree of competence in performances of that kind."[56] It has to be noted that Ryle does not talk about understanding natural or social phenomena, but rather actions performed by other individuals. Consider the following baseball example as an illustration. A spectator who knows nothing about baseball and its rules and has no minimal competence to play baseball will not understand whether the players he observes on the field are playing intelligently or not. He would not be in a position to judge the actions of the teams playing.

According to Ryle, the abilities of appreciating certain performances and executing these performances, like understanding something, are a specific capacity, namely multi-track, and not single-track, dispositions. Recall that these allow for a wide variety of more or less similar practices. Single-track dispositions cover reflexes or habits, meaning that single-track dispositions amount to the same behavior every time it is manifested. Multi-track dispositions, in contrast, refer to capacities of adhering to certain criteria without imposing any or a specific performance that meets the criteria. To this, Ryle adds two important provisos. Namely, that the capacity to execute and appreciate performances does not necessarily involve the ability to articulate criticism about them, and that the ability to perform a specific operation is more demanding than the ability to appreciate it. If this would not be true, there would be no teachers and students who admire their teachers' performances.[57] That is, usually agents first learn how to distinguish excellent from clumsy performances, which requires already some degree of understanding the performance, before learning how to perform excellently themselves.

---

55 Ryle (1949), p. 51.

56 Ibid. p. 53.

57 See ibid. pp. 46, 54f.

Ryle also relates the concepts of understanding, partial understanding and misunderstanding to his notion of knowledge-how. He wants to strengthen his view that understanding is a part of knowing-how even more by comparing the partial nature of knowledge-how and understanding in contrast to (propositional) knowledge. For both understanding and knowing-how it is natural to speak about someone understanding, e.g. chess partially or having partial knowledge of how to play chess. In contrast, it is not possible to speak about the partial knowledge of individual truths or propositions in the same way. Although it is possible and appropriate to talk about partial knowledge about a topic that involves many propositions, it is not possible to say that 'I partially know that today is Monday'. Whereas knowledge can only be imparted, ways of doing can only be inculcated, and inculcation is a gradual process. The gradual nature of knowing-how and of understanding is the source of misunderstanding. Only if someone partially understands Japanese could he misunderstand a text written in Japanese. Therefore, Ryle arrives at the both comforting and motivating conclusion that mistakes are exercises of competence. Without having partial competence, it would not be possible to make mistakes. And importantly, where there are mistakes or misunderstanding, there always is the possibility to correct these mistakes and to gain (a better) understanding. Therefore, learning and participating in controversies is crucial for gaining knowledge-how and understanding as a part of knowing how.[58]

Not only Ryle, but Polanyi, too, is concerned with ideas about understanding, misunderstanding, and sense-making, which he discusses in the context of articulation. Let's have a closer look at Polanyi´s investigation of, as he calls them, articulate and inarticulate intelligence, and how they relate to understanding.

Polanyi starts his analysis with a presentation of three classes of inarticulate intelligence that animals and young children possess as well. These are contriving a skillful action (like a rat learning to depress a lever to receive food), observing a sign-event relation (like a dog learning that the sound of a bell is followed by the appearance of food), and understanding a situation (like a rat that, after learning to run a maze without obstacles, will find the shortest alternative way when some path is blocked). Acquisitions of these types of intelligence are instances of latent learning in which an animal reorganizes its behavior. It exploits a specific means-ends relation to serve some purpose. Understanding a situation is the most elaborate class of inarticulate intelligence because it can be manifested in more numerous and less predictable ways than contriving or observing a situation. The achievement of understanding, the ability to derive various alternative modes of behavior based on the acquired latent knowledge of the situation, represents a basic logical operation that foreshadows the usage of an articulate interpretative framework. In its function as a representation of a complex situation, an articulate interpretative framework

58 See ibid. pp. 57–59.

allows for ever new inferences concerning possible future aspects of the situation with only a minimal exploration of it. Polanyi claims that these three classes of animal learning are three primitive faculties that are more highly developed in humans, meaning that understanding amounts to an act of interpretation when articulation is involved.[59]

When talking about language, Polanyi includes not only writing, but also other forms of symbolic representation in the notion of language, like mathematics, graphs, maps, diagrams, or pictures. This broad notion of language is due to the two principles of language that Polanyi identifies. One principle covers the process of linguistic representation, the other controls the operation of symbols for contributing to the process of thought.[60] Polanyi summarizes the first principle as the application of "the theory of the universe implied by our language to the particulars of which we speak."[61] In his view, this is what humans do when they learn a language, its vocabulary and grammar, and use it to talk about things. Polanyi takes this process to be necessarily unformalized and inarticulate. Learning and applying language to things, denoting things, is a skill in itself. For example, I get to know the English word 'tree' and learn that the word denotes certain objects that have specific features. Once I learned the word 'tree' and the related concept, I will be able to apply it to new objects that fall under the concept, I will be able to call new objects 'tree'. The second principle that Polanyi introduces covers cases of using language for thought. This is an even more demanding ability since it requires the reproduction, storage, transport and re-arrangement of language symbols. A representation of experience needs not only to denote a thing (first principle of language), but rather has to be devised or applied in order to reveal some new aspect of it (second principle of language). For example, if I face an object that I cannot immediately identify as a tree or a bush, I will reorganize the linguistic symbols I possess in order to make sense of that object. Is it a tree, a bush, something else? Polanyi calls the ability to represent experiences in terms of manageable symbols that can be reorganized in order to yield new information the ability of interpretation. However, Polanyi does emphasize that new information is not supplied by merely manipulating symbols. To count as an instance of a real enhancement of intellectual powers by adequate symbols, to gain genuine understanding of an experience or situation, the manipulation of symbols needs to be accompanied by an inarticulate skill of reading the results of the manipulation.[62]

59 See Polanyi (1962 [1958]), pp. 72ff, 76–79.

60 See ibid. pp. 81–84.

61 Ibid. p. 84.

62 See ibid. pp. 84ff.

> [This] performance does require a measure of controlling intelligence. The original situation [...] must be understood and the problem involved in it clearly recognized; then its symbolic representation, including the subsequent operations, has to be correctly performed and the result correctly interpreted. All of this requires intelligence, and it is in the course of these tacit feats of intelligence that the formal operations utilized in the process are accredited and their result accepted by the person carrying them out.[63]

Polanyi specifies how the tacit or personal dimension contributes to the explicit or formal dimension, how language and thought are related, by looking at cases where the two domains fall apart.

> More precisely speaking, we should say that we are referring [...] to a state of mental uneasiness due to the feeling that our tacit thoughts do not agree with our symbolic operations, so that we have to decide on which of the two we should rely and which we should correct in the light of the other. [...] There will always remain certain chances of error – and even of grave error – which arise from our very adoption of an articulate interpretative framework.[64]

That is, the theory of the universe that is implied by any language might be wrong, either completely or with respect to certain aspects of the world. To determine whether a language captures truths about the world, the text (the part of language in question), the conception suggested by it, and the experience on which it might bear have to be considered. Then, three options remain: the language, and thereby the text, is modified, the experience is reinterpreted, or the text is dismissed as meaningless altogether.[65]

Beneath this intellectual strive to establish coherence among language and perception acts an active principle that humans and other animals possess, so Polanyi argues. This principle urges them to discover truths in the world through perception and, in the case of humans, also through language. "We strive for understanding and satisfy our desire for it by seeking to frame conceptions of the greatest possible clarity."[66] Since this striving is already manifested in perception, e.g. when the lens of one's eye gets adjusted by muscles in order to see a certain object sharply, it highlights the important contribution of sense perception to the tacit components of articulated knowledge.[67] "If perception prefigures all our knowing of things, drive satisfaction prefigures all practical skills, and the two are always interwoven. [...]

63 Ibid. p. 87.

64 Ibid. pp. 97f.

65 See ibid. pp. 98f.

66 Ibid. p. 100.

67 See ibid. pp. 100–103.

Therefore, at each of the innumerable points at which our articulation is rooted in our sub-intellectual strivings, or in any inarticulate feats of our intelligence, we rely on tacit performances of our own, the rightness of which we implicitly confirm."[68]

These considerations on an active principle or strive to make sense of the world that, in Polanyi's words, urges all increase in knowledge through perception or language sheds some light on the tacit faculty that enables humans to conciliate experience and language.

> [Humans possess a] power for comprehending a text and the things to which the text refers, within a conception which is the meaning of the text. [...] The urge to understand experience, together with the language referring to experience, is clearly an extension of this primordial striving for intellectual control. The shaping of our conceptions is impelled to move from obscurity to clarity and from incoherence to comprehension, by an intellectual discomfort similar to that by which our eyes are impelled to make clear and coherent the things we see. In both cases, we pick out clues which seem to suggest a context in which they make sense as its subsidiary particulars.[69]

Polanyi repeatedly emphasizes the inextricable relation between articulate and inarticulate intelligence.

> This is the sense in which I called denotation an art. To learn a language or to modify its meaning is a tacit, irreversible, heuristic feat; it is a transformation of our intellectual life, originating in our own desire for greater clarity and coherence, and yet sustained by the hope of coming by it into closer touch with reality. Indeed, any modification of an anticipatory framework, whether conceptual, perceptual or appetitive, is an irreversible heuristic act, which transforms our ways of thinking, seeing and appreciating in the hope of attuning our understanding, perception or sensuality more closely to what is true and right.[70]

Like Ryle, also Polanyi draws a connection to misunderstanding in the context of re-interpreting language. Since our articulate interpretative frameworks will never be immune to inappropriateness and therefore to revision, our understanding of language as well as of the aspects of the world that are denoted will change when language is re-interpreted. When committing verbal mistakes that originate from some inappropriate conception of certain aspects of the world, the subject will feel puzzled and might recognize or even overcome her misunderstanding. Polanyi

68 Ibid. pp. 104f.

69 Ibid. pp. 105f.

70 Ibid. p. 111.

presents an example from chemistry in which re-interpretations allowed for a better understanding. When John Dalton introduced the atomic theory of chemistry in 1808, it was immediately accepted. The atomic theory basically states that all matter is composed of atoms, that all atoms of one element are identical, and that atoms of different elements differ in size and weight. Although chemists used the atomic theory universally, it was not very well understood. Only fifty years later, in 1858, Stanislao Cannizzaro introduced a new articulate interpretative framework in form of a distinction between atomic weight, molecular weight and equivalent weight (weight per valence). These related conceptions have been used interchangeably before, which led to some confusion and conflicts in the chemical community. For example, Dalton rejected Avogadro's Law because it contradicted the atomic theory.[71] As Polanyi states, "the appositeness of Cannizzaro's interpretative framework brought new clarity and coherence into our understanding of chemistry."[72]

Cannizzaro improved the language of chemistry due to his better understanding of the subject matter, which allowed him to develop and use a more appropriate interpretative framework. Polanyi describes the form of confusion, which arose for example in the chemical community in the early 19th century, as a deficiency of intellectual control. Such a deficiency of intellectual control amounts to discomfort and can only be remedied by conceptual and linguistic reform. The divergence of text and meaning, of language and experience, in science but also in everyday life, indicates a problematic state of mind. Every time this is resolved, that is, when the text or experience is re-interpreted or when a text is dismissed as meaningless, some new meaning is created that adheres to standards of clarity and reason. That is, we call a newly discovered kind of beetle a beetle and not a butterfly, because our conception of beetle by which we include the new species makes sense. A modification of our conception of butterfly to cover the new species would not make sense.[73]

Polanyi then relates his discussion of the inarticulate and the articulate manifestations of intelligence to his notions of subsidiary and focal awareness already mentioned throughout section 4.1. While humans pay attention to a specific situation they are concerned with, they subsidiarily adjust conceptions they already possess and change the use of their language so that they can accommodate new things that

---

71 See ibid. pp. 112f. Atomic weight, nowadays called atomic mass, is the mass of atoms of chemical elements. Molecular weight, also known as molecular mass, is the sum of the atomic masses of all the atoms in a molecule. Equivalent mass or, in former times equivalent weight, for chemical elements is the atomic mass divided by the valence. It is the mass of an element which is able to bind or displace one gram of hydrogen. For example, the equivalent mass of oxygen is 16/2 = 8. The valence depends not only on the element, but also on the chemical reaction under consideration.

72 Ibid. p. 113.

73 See ibid. pp. 113–117.

were recognized as new versions of already known kinds of things. The focal attention is targeted at making sense of the situation we are facing, as our inarticulate intelligence adapts and modifies our conceptual framework. This process is comparable to the unconscious interpretation of sensory cues in the context of perception or to the extension of skills by practicing them in as yet unknown situations without being focally aware that one is extending one's skill, so Polanyi argues.[74] The subsidiary search for words to manage a new situation keeps changing the meaning of language. This ability is ultimately manifested in the existence of many different languages, which emerged because different groups of people in different regions of the world at different points in time arrived at different conclusions concerning the conceptions and words they use. The alternative conceptual frameworks sustained by different groups are of course influenced by the things that these groups experience. As a result, the conceptual frameworks express specific theories of the world. Every child accepts the respective theory of the universe implied in a language in the process of learning that language, and every intellectual strive of adults will happen within this framework. One important implication of this according to Polanyi is that humans are divided into groups due to their different vocabularies to interpret the world. This leads to groups that cannot understand each other's way of seeing the world.[75]

In sum, Polanyi argues that any kind of human thinking or reasoning about the world is not possible without language and that we cannot understand any situation or phenomenon in the world without employing our respective language. In his own words:

> Speaking more generally: in order to analyze the use of a descriptive term we must use it for the purpose of contemplating its subject matter, and an analysis of this contemplation will inevitably extend to the contemplated object. It will thus amount to an analysis of the conception by which we are jointly aware both of the term and the subject matter, or more precisely, to an analysis of the particulars covered by this conception: from which we may derive both a more rational use of the term and a better understanding of the things which it designates.[76]

### 4.2.2 Understanding is gradual, multi-track, and its manifestations context-sensitive

What can we take from Ryle and especially Polanyi for a concept of understanding? First, both explicitly state that they take understanding to be a competence, a dis-

---

74 See ibid., p. 118

75 See ibid. p. 118.

76 Ibid. p. 122.

position, or a form of inarticulate intelligence, because understanding comes in degrees. It is natural to speak of understanding something only partially, as it is natural to speak of any competence to be only partially acquired by a subject. This is not the case for (propositional) knowledge. One knows *that p*, or one does not know *that p*, but one does not partially know *that p*. Second, both Ryle and Polanyi expect that understanding can be manifested in various possible and unprecedented ways. To be a multi-track disposition, Ryle demands understanding to meet certain criteria without following any specified procedure. Polanyi takes understanding to be the most elaborated class of inarticulate intelligence, as it enables the generation of alternative and non-predicted results. So, it is not only the case that understanding comes in degrees, that is, that certain experts have a better understanding of their field of expertise in comparison to laypeople, but also that the experts among themselves might have a different understanding of the very same phenomena they are concerned with, since they could understand one and the same phenomenon quite differently from each other. Third, and this point has been stressed by Polanyi specifically, a crucial factor for human understanding is language, an articulate conceptual framework. The languages humans learn and adopt through growing up in specific cultures are the reference frames in respect to which humans understand the world and phenomena in it. Different scientific disciplines or communities, which can be viewed as cultures[77] themselves, developed sophisticated and formalized languages that can accommodate the phenomena these disciplines are concerned with. As different cultures developed different formalizations of the world, different concepts and different languages, they understand the world differently and might not be able to make sense of the conception that another person using a different language applies to certain phenomena.[78] Hence, while experts might understand the same phenomenon differently by using different articulate conceptual frameworks, they will all have manifested the same ability, namely understanding, through aligning language and the phenomenon.

These three aspects, the partial possession or mastering of a competence, the multi-trackness that gives room for various forms of manifestation of the understanding, and the dependence on specific articulated conceptual frameworks, allow for making mistakes or for misunderstanding something. Something like this

---

77 As for example Knorr-Cetina argued, see Knorr-Cetina, K. (1999), *Epistemic Cultures: How the Sciences Make Knowledge*. Cambridge (MA), Harvard University Press, DOI: 10.4159/9780674039681.

78 Thomas Kuhn and Paul Feyerabend had similar thoughts and introduced the idea of incommensurability (of scientific theories) in the 1960s. For an overview see Oberheim, E. & Hoyningen-Huene, P., "The Incommensurability of Scientific Theories", *The Stanford Encyclopedia of Philosophy* (Fall 2018 Edition), Edward N. Zalta (ed.), URL = https://plato.stanford.edu/archives/fall2018/entries/incommensurability/ (last accessed April 16th, 2022). I thank David Lambert for pointing this out to me.

is not possible in the case of knowledge. There is no meaningful way of saying 'I have misknown *that p*', although it is perfectly appropriate to say that I misunderstood *p*. Let me elaborate these three features with some examples.

The first aspect of understanding on which Ryle and Polanyi agree is its gradual nature. In the current debate on understanding among epistemologists and philosophers of science, the gradual nature of understanding is one of the few aspects that is not being challenged by anyone as far as I know. It is not clear, however, how exactly differences in the degree of understanding shall or could be accommodated.[79] Despite the varying attempts to spell out differences in degree of understanding, nobody would oppose that, say, a first grader's understanding of volcanic eruptions is not as good as the understanding a first semester geology student has of volcanic eruptions, which again is probably not as good as the understanding that a geology professor has of volcanic eruptions. I will come back to the graduality of understanding in chapter six in the context of my account of scientific understanding.

The second aspect concerns the feature of understanding being a multi-track disposition. Understanding can be manifested in various unprecedented ways by still adhering to given standards. This idea can be clarified by referring to the second principle of language that Polanyi introduces. To understand a new experience, situation, or phenomenon, an agent needs the ability to manipulate symbols correctly and interpret the result of the manipulation correctly. To correctly manipulate and re-interpret symbols amounts to adhering to the rules of grammar of the respective language and to accommodating the experience. Bringing experience and language into line is an ability. While we are consciously trying to make sense of an experience, we have no access to the ways in which our mind tries to conciliate language and experience. Recall the three options that Polanyi offers for these cases. One could dismiss a text, that part of the language that is targeted towards a specific experience, altogether, as it was the case with phlogiston theory. Alternatively, one could modify the text, which is what happened in the episodes of Cannizzaro's clarification of the atomic weight. Finally, one could re-interpret the experience in light of the existing language, which is what Ludwig Boltzmann did to resolve the specific heat anomaly. Boltzmann, while neither adding nor modifying any concept in the language of physics of his time, used the concept 'degrees of freedom' introduced by James Clerk Maxwell and re-interpreted the behavior of anomalous gases in a way that resolved the specific heat anomaly.[80] In all of these cases, it is not only impossible to determine or prescribe in advance which of the three options should be chosen. Also it is impossible to specify in advance how precisely either of the three

79 See Baumberger, Beisbart & Brun (2017), pp. 26f.

80 For a detailed analysis and discussion of this episode from scientific practice, see de Regt (2017), pp. 205–216.

options will be executed. If the text is dismissed, how will the phenomenon be accommodated? Through a new text and if so, what will that look like? If the text is modified, how is it modified and what will it look like in the end? If the phenomenon or experiences is re-interpreted, what will the new interpretation be? How individuals handle hitherto unknown situations, solve unknown problems, or understand new phenomena can never be known in advance, even if they use the same conceptual framework. The understanding individual herself and the other members of her community can only assess her understanding retrospectively with regard to the grammatical rules of their language and its success in accommodating the experience in question.

Concerning the third aspect, the possibility of misunderstanding due to modified or different conceptual frameworks, every conceptual framework might be wrong in fundamental ways, as Polanyi recognizes. Humans construct conceptual frameworks because they strive to discover truths about the world, and through the storage of a great amount of information in language, more and more details of the world can be recognized, analyzed, and, ultimately, understood. Concepts that persist over a long period of time have proven to be successful in many instances. However, no concept will ever be immune to revision. Languages improve, concepts are changed or added to accommodate new experiences better.[81] These changes in language reflect the changes, and possibly even the degrees, of understanding. For example, the language of phlogiston theory has been abandoned by chemists. From our contemporary perspective, chemists who understood combustion through phlogiston theory completely misunderstood the phenomenon, since we know nowadays that phlogiston does not exist. However, whether an individual understood or misunderstood is a matter of context. Phlogiston theory was very successful for almost 75 years and its proponents had some good empirical justification to apply and to defend it. They had reasons to assume that phlogiston exists and that the theory brought them closer to reality. When phlogiston theory has been developed, no one could know that it is false. This had to be discovered

81 This idea is comparable to the concept of epistemic iteration that Hasok Chang introduced. Based on his discussion of the development of thermometry, Chang argues that ""epistemic iteration is a process in which successive stages of knowledge, each building on the preceding one, are created in order to enhance the achievement of certain epistemic goals. ... In each step, the later stage is based on the earlier stage, but cannot be deduced from it in any straightforward sense. Each link is based on the principle of respect and the imperative of progress, and the whole chain exhibits innovative progress within a continuous tradition." Iteration provides a key to understanding how knowledge can improve without the aid of an indubitable foundation. What we have is a process in which we throw very imperfect ingredients together and manufacture something just a bit less imperfect." Chang, H. (2004), *Inventing Temperature: Measurement and Scientific Progress*, New York, Oxford University Press, DOI: 10.1093/0195171276.001.0001, p. 226.

over time.[82] The language of modern chemistry is different than its predecessors in the 17th and 18th century, and this change in language accompanies and mirrors the gradual nature of understanding.

Episodes of revised conceptual frameworks and the accompanying understanding can, of course, be found in other disciplines, too. Consider, for example, de Regt's case study of early quantum physics and the rivalry between Erwin Schrödinger and Werner Heisenberg regarding the construction of an adequate theory of the structure of atoms. Schrödinger's wave mechanics represented the atomic structure in terms of wave functions, which was a very different form of articulation compared to Heisenberg's theory of matrix mechanics. Matrix mechanics did not represent atomic structure directly, but rather described relations between observable quantities like frequency or intensities of spectral lines. Additionally, Heisenberg formulated his theory in the mathematical language of matrices, with which most physicists were not familiar in the early twentieth century. That is, Heisenberg used a different language, a different articulate interpretative framework, than Schrödinger, while both were concerned with the same phenomenon, namely, the atomic structure. As a result, they judged each other's theory as unintelligible. Also, both theories were not without problems in accommodating known atomic phenomena, which again encouraged further critique by both parties against the opposing camp. It was Wolfgang Pauli, a companion of Heisenberg and supporter of his theory, who claimed that the new theory, or conceptual system, of matrix mechanics first has to be learned by everyone in the physical community, that the conceptions employed in the new theory must be understood, before it can be successfully used. Ultimately, the work of Schrödinger and Heisenberg has been combined and resulted in the theory of quantum mechanics that is known and taught today.[83]

### 4.2.3 Understanding is the ability to make sense of a phenomenon

Given the fundamental difference between knowing-how and knowledge-that, in this section I aim to show that understanding is a kind of knowing-how because, in alignment with Ryle and Polanyi, understanding is a gradual and multi-track disposition whose manifestations are context-sensitive. The concept of propositional knowledge as being something like justified true belief cannot capture what we associate with understanding, namely some competence of making sense of a situation, create a meaning, or yielding new information. How does this notion of understand-

82 For a detailed discussion of the merits of phlogiston theory, see Chang, H. (2012), *Is Water $H_2O$? Evidence, Realism and Pluralism*, Dordrecht, Springer, DOI: 10.1007/978-94-007-3932-1.

83 For an extensive and detailed discussion of this case study from physics, see De Regt (2017), chapter 7.

ing fit with my definition of abilities developed in chapter 4.1? There, I presented the following definition:

*x* is an ability if and only if *x*

i. is a disposition to perform a cognitive or physical activity successfully with respect to relevant standards,
ii. has been learned and trained in a specific social context, and
iii. manifests in processes that are partially tacit (i.e. that can never be made fully explicit).

The notion of understanding sits well with this. First, understanding is a disposition to perform a cognitive activity, namely making sense of a phenomenon through aligning language with experience. The match of language and experience must adhere to the non-formalizable standards upheld by the respective community without consciously reflecting on them. Second, those implicit standards are learned through participation in a community. By being raised and trained in a language, in a cultural community, humans learn how to use and speak the language, and to modify it to accommodate or manage new experiences. However, not just any modification or manipulation of symbols is allowed. Every language prescribes rules about its use that have been implemented due to their past success. Any modification of language must be plausible in light of the rules of grammar and of the experience is shall accommodate. The interpretation arrived at by the individual's understanding must make sense in light of the language, the theory of the universe, and the empirical evidence. Since the development of language serves the human striving of arriving at an ever more precise and accurate comprehension of the world, the rules that guide the use of a language serves this superior goal as well. Therefore, every member of a community agrees to respect the rules of a language, internalize and act upon them. However, the rules themselves are not valid eternally, but rather susceptible to change if they cannot accommodate (some) experience at all. The rules that govern the use and modification of a language depend on the specific context of the community. Hence, these rules can only be learned and internalized if one is a member of the respective community. Humans learn how to bring language and experience into accordance with one another by interacting with other, already versed individuals. Novices, both in science or in everyday life, learn and train how to understand phenomena under the guidance of teachers and supervisors. Teachers and supervisors assess whether the understanding of a phenomenon that students acquired is adequate. When the students demonstrated that they arrived at an adequate understanding often enough, it will be determined that they successfully acquired the ability to understand particular phenomena in the world. And third, the manifestation of understanding is partially tacit for the subject, may it be a student or any other, more experienced person. While the person is consciously aware of the phenomenon, the situation, or experience she wants to understand, she is not aware of all her mental performances through which she ultimately conciliates experience and language.

After arguing in section 4.1 that abilities are dispositions to succeed, I suggested throughout this section that understanding could be conceptualized as an ability, a disposition to succeed in making sense of some phenomenon or experience. Whenever a subject understood something, she will have manifested the ability to understand by creating some meaning of the phenomenon that is acceptable given the respective standards. But how exactly do subjects do this, how exactly is understanding manifested? I provide an answer to this question in the next section.

## 4.3 The manifestation of scientific understanding

In the previous two sections, I developed and defended a definition of ability and argued that understanding fits this definition of ability. Now, if we accept this and take understanding to be an ability, a disposition to succeed, how exactly is understanding manifested? I already claimed that understanding manifests in aligning language with experience, but can this be spelled out in more detail? That is the task to which I shall now turn. In section 4.3.1, I address a prominent concept in the literature on understanding, grasping, and clarify what I mean with this notion. Following that, I argue why, in addition to grasping, articulating an explanation is necessary for (scientific) understanding in section 4.3.2 by drawing on Michael Polanyi's work on articulation and Mark Newman's model of understanding. This also is a clarification of the necessary relation between scientific understanding and scientific explanation, for which I argued in chapter three. In sum, I argue in this section that scientific understanding is manifested in the process of grasping relation of a phenomenon and articulating these relations as explanations. Importantly, my claims in this section are intended to cover scientific understanding, understanding gained in science, and not necessarily all kinds of non-scientific understanding. Neither do I claim that all kinds of understanding are manifested through grasping relations and articulating explanations, nor that only scientific understanding is manifested through grasping relations and articulating explanations. However, an analysis of a categorization of types of understanding that are manifested in this way and those that are not is a topic for further research and not covered by this book.

### 4.3.1 Grasping relations

When we consider possible manifestations of the ability to understand, one candidate that can be considered is grasping. The notion of grasping is ubiquitous in the debate on understanding and closely related to discussions about abilities in the context of understanding. Quite many scholars try to clarify what understanding is by reference to the notion of grasping. To my knowledge, Jonathan Kvanvig, whose view I presented in detail in section 3.3, was the first who gave grasping a prominent role in his analysis of understanding. According to him, "to understand is to grasp the variety of [...] connections [between pieces of information]."[84] Unfortunately, it remains unclear what Kvanvig means exactly when he talks about grasping, as he does not elaborate this term any further. If "grasping" is used merely as another term for "understanding" without any further explication, the mere introduction of this term will not lead to any insights about understanding. Hence, several different and partly conflicting accounts of grasping are offered by various scholars.[85]

Before I address different views of grasping, let me emphasize the one common and basic assumption that unites all the different views of grasping. Whatever grasping might be in the end, grasping is taken to demarcate understanding from knowledge. Baumberger, Beisbart & Brun provide a nice example that elucidates this basic idea:

> Suppose that a climate scientist explains to her young son that the global mean surface temperature has massively increased since the middle of the 20th century because of increasing greenhouse gas concentrations. Since she is right and her son has good reasons to believe her explanation, he may be said to know why the global mean temperature has increased. But he does not seem to understand why. When asked why this is so, all he can do is to repeat his mother's explanation. The problem seems to be that he does not really grasp the explanation. But what exactly is he lacking?[86]

It is the answer to the question raised at the end of the example on which scholars working on understanding disagree, but they do not disagree on the problem. Here, I present two different interpretations of the notion of 'grasping', the "naturalistic view" and the, to put a label on it, "grasping as abilities"-view. I will argue that the "naturalistic view" of grasping is more plausible.

A basic insight of naturalists is provided by Daniela Bailer-Jones, namely that "understanding has a subjective component, in addition to the publicly accessible

---

84 Kvanvig (2009), p. 96.

85 For a good overview of different accounts of and controversies concerning grasping, see Baumberger, Beisbart & Brun (2017), pp. 12–17.

86 Ibid. p. 12.

component represented by explanation, in the sense that understanding takes place in an individual's mind."[87] Michael Strevens thinks along similar lines. He takes grasping of a correct scientific explanation of a phenomenon to be necessary for understanding this phenomenon and views grasping as a "fundamental relation between mind and world, in virtue of which the mind has whatever familiarity it does with the way the world is."[88] On that basis, Reutlinger et al. conclude that grasping is a philosophically primitive notion, i.e. that it does not matter for a philosophical analysis of understanding that grasping cannot be clearly defined. Since they take this to be a task for cognitive scientists, they call their view of grasping the "naturalistic view". However, what is important for philosophical accounts of understanding is that grasping is the subjective component of understanding, that grasping allows for some epistemic accessibility of a phenomenon for scientists, or subjects more generally. Grasping is taken to be a fundamental relation between mind and world.[89] Therefore, grasping (having epistemic access to) a phenomenon is a necessary condition for understanding it, but grasping is not identical to understanding.

However, this is not the only view on grasping. I call an alternative conception the "grasping as abilities"-view. Christoph Baumberger argues that "grasping the causes or reasons why p [...] is better spelled out in terms of having certain abilities that are not required for simply *believing* that the factors in question are the causes or reasons why p."[90] The possession of knowledge depends on certain abilities, too, like memorizing and quoting information, but these are not the abilities that are necessary for understanding.

---

87 Bailer-Jones, D. (1997), *Scientific Models: A Cognitive Approach with an Application in Astrophysics*. Ph.D. Thesis, University of Cambridge, p. 122.

88 Strevens, M. (2013), "No Understanding without Explanation", *Studies in History and Philosophy of Science A*, 44 (3), pp. 510–515, DOI: 10.1016/j.shpsa.2012.12.005, p. 511. It should be noted that Strevens adopts an ontic conception of explanation. That is, he views explanations as physical entities that exist in the causal structure of the world. The ontic conception of explanation is opposed to the epistemic conception of explanation, according to which explanations are representations of phenomena in the physical world. So for Strevens, grasping a scientific explanation means that a subject grasps an actual causal process in the world, and not merely a representation of that process. See Strevens (2008), p. 6. In contrast to Strevens, I endorse an epistemic conception of explanation. This fundamental difference aside, Strevens and I agree on a basic notion of grasping in the sense that (aspects of) phenomena in the world have to be grasped, and not merely an explanation in the sense of a representation, if a subjects wants to gain understanding of the phenomena in question.

89 See Reutlinger, A., Hangleiter, D. & Hartmann, S. (2018), „Understanding (with) Toy Models." *British Journal for the Philosophy of Science*, 69 (4), pp. 1069–1099, DOI: 10.1093/bjps/axx005, pp. 1082–1085.

90 Baumberger (2011), p. 73.

Baumberger suggests that for having understanding "why p (where q is why p) then [one is] (to some extent) able

> i) to comprehend and render an explanation of p which shows (e.g. by means of a generalization) how p depends on q,
> ii) draw the conclusion that p (or that probably p) from the information that q, and
> iii) for some p* and q*, similar but not identical to p and q, draw the conclusion that p* (or probably that p*) from the counterfactual assumption that q*, and, counterfactually assuming that p*, explain it with the help of q*."[91]

If the aspired understanding is not limited to grasping the causes or reasons for a phenomenon, the grasping of different or more dependency relations amounts to more of the same abilities that are already necessary for understanding the causes of a phenomenon, according to Baumberger.[92] Steven Grimm adopts a similar view on grasping as Baumberger and describes grasping dependency relations as "being able to "see" or anticipate how varying the value of one of the variables will lead (or fail to lead) to a change in the value of another variable. What this grasp involves is thus the ability to make modal inferences or to "see" into modal space."[93] Both Baumberger and Grimm view grasping to encompass other and possibly several different reasoning skills. For Grimm, grasping involves (at least) the ability to make modal inferences, whereas Baumberger takes grasping to include comprehending and providing an explanation as well as making varying kinds of inferences, as I state above.

What shall we make of these two different conceptions of grasping? I argue that the naturalistic view of grasping is more plausible than the "grasping as abilities"-view. Bailer-Jones correctly points out that understanding takes place in the minds of individuals and I argued in section 4.2 that understanding is taken to be an ability to make sense of a situation, create a meaning, or yield new information. Hence, it is plausible to conceive understanding as a cognitive ability that is manifested in our minds. Therefore, the manifestation of understanding is a cognitive process as well. However, as we often wish to understand phenomena, things, or situations that take place outside of our minds, in the world, we need to establish some connection between our minds and the things we want to understand. If we do not do this, I would not know how it should be possible to understand anything outside of our minds. Here grasping comes in. For an account of understanding, it is sufficient to think of grasping as getting epistemic access to a phenomenon. The metaphors of 'seeing, recognizing, or becoming aware of' relations of phenomena in the world might be instructive here. This view is in accordance with Strevens and Reutlinger et

91 Ibid. p. 73.

92 See ibid. p. 79.

93 Grimm (2017), pp. 216f.

al., who identify grasping as having epistemic access to the object of understanding. Hence, I am also taking a naturalistic stance on grasping. If we want to understand phenomena in the world, we need access to them. When a person grasps a relation, this relation somehow catches the attention of the person, it gets into her focus. She somehow recognizes that there is something interesting or relevant about the phenomenon that she wants to understand. And it only happens in the next step, after recognizing that some relation is there, that the person applies modal, counterfactual, inductive, deductive or analogue reasoning to make sense of the relation that has just been grasped. I take grasping to be a process that precedes and is distinct from other reasoning processes. I do not view grasping to be a composition of reasoning skills, as Baumberger and Grimm argue, because a person cannot make modal inferences or reason about something that she is not aware of at all. Individuals first need to grasp something they should or could reason about, before they can actually reason about it. Without establishing a relation between mind and world, without grasping, no reasoning about things could ever begin. Hence, grasping is worth to be taken as a distinct process.

In sum, grasping is the process of getting epistemic access to relations of the phenomenon that shall be understood. But what about explanation? I argue in chapter three that understanding requires explanation, but my conception of grasping as 'seeing' relations does not capture or include explanation. This problem can be solved in taking grasping to be only a partial manifestation of understanding. The complete manifestation of understanding requires grasping relations of the phenomenon as well as articulating explanations of the phenomenon. I address the second component of the manifestation of understanding in the next section.

### 4.3.2 Articulating explanations

How to flesh out the idea that articulating explanations is a necessary component of the manifestation of understanding? Michael Polanyi, again, provides helpful insights on this matter, when he investigates the role of articulation for scientific thought.

Humans rely on articulate interpretative frameworks as representations of complex situations to assist and guide their actions. Applying an articulate interpretative framework to a situation decreases the amount of mental work that a subject has to invest for analyzing a situation. Humans do not need to explore any new situation in all its complexity because the articulate framework already provides an interpretation of the situation, so Polanyi argues.[94] Therefore, the subject can almost immediately pass on to solve a specific problem in the given situation, without spending much time and energy to make sense of the situation in the first place.

94 See Polanyi (1962 [1958]), p. 76.

> Moreover, by being prepared to speak in our language on future occasions, we anticipate its applicability to future experiences, which we expect to be identifiable in terms of the natural classes accredited by our language. These expectations form a theory of the universe, which we keep testing continuously as we go on talking about things. So long as we feel that our language classifies things well, we remain satisfied that it is right and we continue to accept the theory of the universe implied in our language as true.[95]

The ability of using language in thought enhances the intellectual powers that humans possess. And Polanyi identifies several levels of articulation. The highly specialized scientific nomenclatures, symbolic operators or numerical denotations are expansions of ordinary speech that enable scientists to master even more complex situation or problems. Articulation enables systematization and manageable records that assist memory as well as speculative imagination because the crucial aspects of any situation can be presented in a comprised form through articulation.[96] "Articulation pictures the essentials of a situation on a reduced scale, which lends itself more easily to imaginative manipulation than the ungainly original."[97]

This 'theory of the universe' is already implied in the sense perceptions, according to Polanyi. Perception serves animals as well as humans to find their way around in the world, to find food or avoid threats. That is, perception provides us with the clues that we need to solve problems we are confronted with in everyday life. We trust our perception and experience to convey the things in the world to us in the way the things really are. Perception already establishes a 'theory of the universe'. Using language and applying words to objects or situations that we have already identified through our perception is an extension of that 'theory of the universe' that we already possess. Through language, it is possible to develop clearer and less ambiguous conceptions of the universe, and the various objects it comprises.[98]

Verbal and other linguistic pointers aid and enhance our mastery of any issue we are confronted with because they enable us to manage massive amounts of experiences and information. However, Polanyi does not argue that we stick to any articulate framework forever after we had learned it. On the contrary, he analyses how language is susceptible to re-interpretation.[99] Since the world and our experience of it are constantly changing, the meaning of language and the conceptual framework we are applying in a specific situation will also be modified with every new instance in which it is applied. We seek to achieve more clarity and precision in our

95 Ibid. p. 83.

96 See ibid. pp. 86ff.

97 Ibid. p. 88.

98 See ibid. p. 100ff.

99 I mentioned this point, too, in section 4.2 and illustrated it with examples of new articulated frameworks in chemistry and physics.

language as well as in our experience in order to find solutions for problems we are confronted with. Any modification or re-interpretation of an articulated conceptual framework is done owing to our hope of getting closer to reality. And modifying, re-interpreting, or even learning a language in the first place does change our ways of thinking. Polanyi holds that the modification of a conceptual framework is a subsidiary process that takes place while we focus on the situation we are dealing with. If a conceptual change proves to be successful, it will get established in the framework and transmitted to other member of the community who will continue to use the modified conceptual framework.[100]

In short, Polanyi argues that the use of language rendered the intellectual achievements of humankind possible. Humans are able do deal successfully with complex problems or situations, because our knowledge of the universe is stored, presented and used in the form of articulated language. Language assists thought; it enables sophisticated thought processes that would not be possible without the use of language as a guiding interpretative framework. This is the reason why humans, some of which are scientists, cannot make sense of phenomena without using language. Humans are driven by an "urge to understand experience, together with the language referring to experience. [...] While our thoughts are of things and not of language, we are aware of language in all thinking [...] and can neither have these thoughts without language, nor understand language without understanding the things to which we attend in such thoughts."[101] That is, we cannot understand anything in the world without thinking in our respective language, and scientists cannot reason about newly discovered relations or aspects of the world without using language.

A very similar line of thought can be found in the work of Mark Newman, who provides an example that illustrates the difference between *knowing an explanation* and *understanding the phenomenon that is presented by this explanation*:

> Muons [...], which have a proper lifetime of only $2.2 \times 10^{-6}$ seconds, can last the longer travel time of $333 \times 10^{-6}$ seconds as they traverse from the upper atmosphere to the earth's surface. How is this possible?
> Explanation $e_{muon}$: muons are elementary particles which travel at 0.999978 times the speed of light. Entities that travel this fast are subject to the time dilation effect of Special Relativity. Time dilation is given by the relation: $\Delta t = \Delta t_0 / 1 - (u^2/c^2)$. Where $\Delta t$ is change in time in Earth's reference frame, $\Delta t_0$ is change in the proper time of the muon, $u$ is the muon's speed and $c$ is the speed of light. Doing the cal-

100 See Polanyi (1962 [1958]), pp. 108–111, 118–123.

101 Ibid. p. 106

culations we find that although it initially seems impossible, muons can actually last long enough to survive the journey.[102]

A competent English speaker with working memory who reads this explanation and comes to believe it will *know* this explanation of the lifespan of muons. But simply knowing, remembering and even re-stating or re-formulating this explanation in a linguistic sense does not amount to *understanding* the phenomenon of the lifespan of muons. What is necessary for gaining understanding of muons with this explanation is knowledge of the meaning of the concepts that are included in the explanation, not merely understanding the propositions linguistically. Only if one knows what a reference frame, proper time, and the time dilation equation *mean* can one really understand the lifespan of muons. According to Newman, knowledge of an explanation can be identified with a linguistic understanding of an explanation, whereas understanding of the phenomenon, which is presented by an explanation, requires conceptual knowledge of the 'deep meanings', as he calls it, of the concepts involved.[103]

How is it possible to achieve this conceptual knowledge of the deep meanings of concepts? Newman offers a first possible answer by referring to Conee and Feldman[104] and proposes that background beliefs are necessary in order to really understand an explanation. An expert, in contrast to a novice, has a robust set of background beliefs concerning the concepts used in an explanation. Therefore, the expert is in a position to categorize these concepts and make sense of the explanation as a whole. However, having the relevant background beliefs, i.e. knowing the meanings of the concepts, is not sufficient for an account of understanding the phenomenon. Additionally, the expert also has to appropriately *use* these background beliefs. If she does not do so, her understanding, if we attribute some understanding at all, will only be a result of lucky guessing and she will not be justified in accepting her understanding.[105] Importantly, abilities play a crucial role in Newman's model as well and his demand that relevant background beliefs must be possessed as well as used to understand a phenomenon is in line with Polanyi's analysis of the necessary interplay between articulate and inarticulate intelligence in the process of making sense of, understand, phenomena in the world.

---

102 Newman, M. (2017), "An Evidentialist Account of Explanatory Understanding." In Grimm, S. R., Baumberger, C. & Ammon, S. (eds.), *Explaining Understanding. New Perspectives from Epistemology and Philosophy of Science*, pp. 190–211, New York and London, Routledge, pp. 192f.

103 See ibid. p. 193. Gilbert Ryle formulated and investigated a similar question: "What [...] is the difference between hearing what a speaker says and making sense of what he is heard to say?" Ryle (1949), p. 51.

104 See Conee, E. & Feldman, R. (2004), *Evidentialism: Essays in Epistemology*. Oxford, Oxford University Press, DOI: 10.1093/0199253722.001.0001.

105 See Newman (2017), pp. 193ff.

Newman's model of understanding, which he calls the "Inferential Model of Understanding" (IMU), includes the following concepts:

> (K) Knowledge of an explanation is an accurate, justified representation of the explanation's propositional content.
> (U) Understanding an explanation is achieved when the representation of an explanation's propositional content is internally connected by correct inferences.
> (UT) S understands scientific theory T iff S can reliably use principles $P_n$ constitutive of T to make goal-conducive inferences for each step in a problem-solving cycle, which reliably results in solutions to qualitative problems relevant to that theory.[106]

My focus lies on (K) and (U). Newman identifies knowledge of an explanation (K) with having linguistic understanding of the explanation. If this is achieved, the subject will have grasped[107] the meanings of each proposition that is involved in the explanation. She will be able to represent the explanation that reflects this grasping. If a person has explanatory understanding (U), she will have linguistic understanding and, *additionally*, link the explanation with correct inferences by exercising default reasoning. This is an implicit, tacit process. Newman takes the concept of default reasoning from the work on mental models in cognitive psychology. The basic idea is that our mental representations are built on rules. These rules are stimulated by default expectations that we take to be correct as long as we have no counterevidence. For example, when we see a black cat, we activate a set of rules that constitute the concept "black cat" and we stick to this mental representation until we gain perceptual evidence that, in fact, what we are seeing is a small black dog instead. The first implicit reasoning process, the expectation and rule-activation, is what Newman calls "default reasoning". This form of reasoning is said to govern most of our everyday reasoning.[108]

The abilities that keep K and U apart are correct inferences by using default reasoning. They are a form of knowing-how, according to Newman. Since he treats knowing-that and knowing-how as non-reducible, a view that accords with the discussion I present in section 4.1, these abilities are not a form of propositional knowledge. In order to gain (U), we need to recognize the appropriate relations as well as the relata presented by the explanation. A necessary prerequisite to do so is to know generative relations like "allowed", "caused", "created", "forced" or "generated". For

106 Ibid. p. 199. I am mentioning UT here for the sake of completeness, but I will not address it in further detail.

107 Newman uses the term 'grasping' without clarifying what he means with it. I use Newman's terminology here, although his notion of grasping might not be identical to my notion of grasping that I present in the previous section.

108 See ibid. p. 200.

every explanation, the correct generative relations between the explanandum and (parts of) the explanans have to be chosen. One single explanation might include several different generative relations and for every step in the explanation, the correct relation has to be used. If a person fails to do so, she will not have understood the explanation. Take again the explanation of the lifespan of muons. Amongst other things, it is stated in the explanation that "Time dilation is given by the relation: $\Delta t = \Delta t_O / 1 - (u^2/c^2)$." A novice will not understand the explanation if she takes this relation to be causal. It is not the case that a change in the earth's relative time causes a change in proper time. Rather, the relation presented here is a sufficient condition and this has to be recognized by the novice. Therefore, the abilities to know and select the appropriate generative relations are an essential difference between knowing and understanding an explanation.[109]

However, this is not the whole story yet. Newman goes one step further and argues that, in addition to recognizing generative relations, an explanation schema for these relations has to be articulated. An explanation schema is a type of cognitive structure that is defined by a set of generative relations. Take the two generative relations "eating generates growth" and "greater size generates slower movement", which are used to articulate the following explanatory schema to explain the extinction of the dinosaurs: "The dinosaurs ate a lot which caused them to grow enormously, which slowed their escape from predators, which caused their extinction." This is an example of an explanatory schema, although an incorrect one. In this case, some appropriate generative relations have been recognized, but not all of them are appropriate. As a result, the person who constructed and articulated this explanatory schema does not understand the extinction of the dinosaurs because she failed to infer the correct generative relation between explanans and explanandum.[110]

The difference between knowing an explanation and understanding it presented by the IMU model of understanding is summarized by Newman as follows:

> IMU adopts the idea that explanatory understanding (U) surpasses the cognitive achievement of knowledge (K) in virtue of the subject activating not only appropriate generative relations from memory, but also articulating those relations in the correct explanatory schema. Without these skills we may come to understand linguistically what is being said, but fail to insert the appropriate relations or relata, and hence fail to explanatorily understand.[111]

The work from Polanyi as well as from Newman suggest that the articulation of an explanation is the second part of the manifestation of understanding, which follows

109 See ibid. pp. 202f.

110 See ibid. p. 203.

111 Ibid. p. 203.

the grasping of relations of the phenomenon. As I take grasping to be a process that is distinct and prior to other reasoning processes, I also demarcate articulating an explanation from other reasoning processes. I do so because the possession and successful manifestation of various reasoning abilities does not automatically amount to the articulation of an explanation. The case of James Clerk Maxwell who tried to make sense of the specific heat anomaly, which I mentioned already in section 4.2.2, is a good example to illustrate this point. Maxwell possessed impressive reasoning and calculating skills. He even introduced a completely new concept, the 'degrees of freedom', in order to make sense of the phenomenon. He realized that the available language of physics cannot accommodate the specific heat anomaly, which is why he introduced the concept 'degree of freedom'. And the concept 'degree of freedom' would be meaningless if Maxwell could not relate it to something in the physical world, in this case, the kinds of motions of molecules. However, despite all his efforts and accomplishments, Maxwell was not able to articulate an explanation of the specific heat anomaly, although he grasped that the phenomenon has something to do with the kinds of motion that the gas molecules exhibit and spent years of his life thinking about and trying to solve the issue. Maxwell contributed groundbreaking achievements to physics, like his classical theory of electromagnetic radiation and his equations for electromagnetism, which sufficiently proof his exceptional reasoning skills, but the specific heat anomaly remained a mystery to him.

In contrast to Maxwell, Boltzmann was able to make use of the available concepts and articulated an explanation of the specific heat anomaly through his dumbbell model. Boltzmann's success might have been due to the extended articulated conceptual framework, which included the concept 'degrees of freedom', that neither Clausius nor Maxwell had at their disposal, at least not from the beginning of their investigations. In Polanyi's view, Boltzmann would never have had the thought processes that ultimately led him to the development of his dumbbell model if he had not had the concept 'degrees of freedom' at his disposal. Without this concept, Boltzmann could not have reasoned with it and could not have formulated an explanation of the specific heat anomaly in terms of degrees of freedom. As Polanyi argues, scientists learn the sophisticated and specialized language of their discipline during their education. That is, scientists think about the phenomena they try to understand in terms of the specific language that they learned. And this language might change when deficiencies are recognized, as in the case of the research on the specific heat anomaly, where the articulate conceptual framework was extended by Maxwell to include the concept 'degrees of freedom'. And in Newman's account, Boltzmann would never have identified the relation between the specific heat anomaly and the degrees of freedom, if the concept 'degrees of freedom' did not exist and, hence, could not be a candidate for a relatum.

Grasping relations of phenomena in the world and articulating them in form of explanations are the two components of the manifestation of (scientific) understanding. Importantly, in my usage the term 'articulating' refers to the construction of an explanation in an individual's mind, and not to any form of expressing or communicating. An individual cannot express or communicate any explanation if she has not articulated this explanation in her mind beforehand. However, I do think that as soon as a subject articulated an explanation in her mind, she will be able to communicate this explanation in some way to other subjects. And since instances of understanding are grounded in a fundamental principle urging subjects to discover truths about the world, subjects who understood something will also want to communicate the explanation she articulated. By making an articulated explanation publicly accessible, her understanding is publicly accessible and can be scrutinized by other subjects. In doing so, a subject can get additional justification and support that she understood something correctly, that she discovered some truth.

### 4.3.3 The manifestation of understanding

To sum up, I argued in this section that grasping relations of a phenomenon and articulating these relations in form of explanations are the manifestation of scientific understanding. Together, grasping and articulating manifest understanding. Grasping denotes the process of gaining epistemic access to the phenomenon that shall be understood, the process of 'seeing' or 'recognizing' some relation of the phenomenon. Subsequently, the subject that grasped some relation of the phenomenon will resort to the conceptual framework she uses in order to represent the grasped relation in form of an explanation.

Since understanding manifests in the process of grasping and articulating explanations, understanding a phenomenon is a procedural ability. The procedural manifestation of understanding is partially tacit for the subject. She will not be able to explicitly state how exactly she gained understanding of a phenomenon, why or how certain observations or data caught here attention, how she grasped a relation she did not know before and how she articulated an explanation of the grasped aspect of the phenomenon by using the specific language she possesses. The ability to understand phenomena is an instance of inarticulate intelligence. However, when a subject gained understanding of the phenomenon, she will be able to make explicit what she understood. That is, she will be able to express the explanation she articulated since the manifestation of understanding relied on the vocabulary of her language. If a person looks at an orrery, she might gain understanding of the planetary motion in our solar system without having explicit access to how exactly she was able to grasp information represented by the orrery. But once she gained understanding of the apparent retrograde motion of mars, to return to an example from Peter Lipton discussed in section 3.1, she can express and communicate what she understood since she thought about the represented phenomenon in the vocabulary of her language. That is an instance of articulate intelligence, the construction of an explanation, which is grounded in inarticulate intelligence, in understanding.

## 4.4 The inextricable relation between understanding and knowledge

What is understanding? Is understanding an ability or a type of propositional knowledge? And if understanding is an ability, how is it manifested? Those were the questions I set out to answer in this chapter. I started with an analysis of the concept 'ability' and developed a definition of ability as dispositions to perform a cognitive or physical activity successfully with respect to certain relevant standards, which have been learned and trained in a specific social context and whose manifestations are partially tacit. This definition of ability accommodates performed activities of subjects that are often or usually labelled skillful, for example athletic or artistic performances, and also theoretical activities like logical reasoning or calculating. I then argued that understanding itself should be regarded as an ability to make sense of a phenomenon, a situation, or an experience, and that such a conception of understanding does not conflict with my argument developed in chapter three that understanding requires explanation. The process that manifests understanding consists of two partial processes, namely grasping relations of the phenomenon that one tries to understand and articulating the grasped relations in form of an explanation.

Why should understanding be viewed as an ability and not as a form of propositional knowledge? Because one and the same phenomenon or experience can be understood in various different ways, using different languages and arriving at different interpretations. Ptolemy understood the motion of heavenly bodies differently than Copernicus, Lavoisier understood combustion differently than proponents of phlogiston theory, and Schrödinger and Heisenberg understood atomic structures differently before they integrated their languages to arrive at a more precise interpretation. All of these individuals have two features in common. First, all of them understood the phenomenon they wanted to understand. They arrived at an interpretation of the phenomenon that accommodated the language they used and the worldly situation they had access to, they were able to make sense of the phenomenon. Second, all the mentioned individuals were striving for truth. All did their best in light of their resources, their language and experience, that they had at their disposal to discover truths about the world. That Ptolemy was wrong in seeing the earth at the center of the universe could only become apparent when our language and our experience of the world developed and with the help of more sophisticated instruments or measurement devices. Whether someone understood or misunderstood a phenomenon can only be assessed in light of a specific context but the ways in which a phenomenon can be understood, how language and phenomenon are conciliated, are countless and cannot be explicitly articulated or predicted in advance. This is different for knowledge. Either ones knows *that p* or one does not, and *that p* can be explicitly stated. Knowledge is not gradual, multi-track, and context-sensitive in the way understanding is. Merely possessing knowledge does not enable a subject to master and combine her language and the phenomenon to which it is applied. Knowledge, in its classical formulation, is justified true belief. A belief is something completely different than the demanding activity of understanding. This differentiation between understanding and knowledge does not only fit Ryle's distinction between knowledge-how and knowledge-that, with which I started this chapter, but also the view from virtue epistemologists that knowledge is an intellectual success achieved through ability.

Understanding is an ability to make sense of experiences, situations, or phenomena in the world, to solve arising puzzles concerning them. Understanding is the ability to generate new knowledge, knowledge that captures the interpretation of an experience that an individual made. We cannot articulate or communicate the understanding itself, that is, how we managed to grasp a relation and articulate this relation in an explanation because it is an ability. But what we can, do, and sometimes even should articulate and communicate, is the result of our understanding, the phenomenon that we have understood, and the interpretation of the phenomenon we arrived at through understanding. In order to understand something that lies outside of our minds, we need to get some access to the thing we want to understand. This happens through the process of grasping relations of the phenomenon or situation. Grasping can be described as recognizing a relation that might have something to do with the thing we want to understand. However, merely grasping relations in the world is not sufficient for understanding because grasping, in the way I conceptualize it in section 4.3.1, does not entail the ability to make sense of what has been grasped. This happens through the articulation of the grasped relation in the vocabulary of a language. Thus, understanding requires grasping relations as well as articulating explanations. Once we arrived at an interpretation of a phenomenon, once we managed to bring our experience and our language into accordance, we arrive at the belief that our interpretation represents a true aspect of the world. We arrived at a justified, possibly true, belief. We generated knowledge. However, holding a justified true belief is something completely different than the ability to grasp relations in the world and articulate them in explanations, that is, to conciliate experience and language and generating new knowledge by ourselves.

Why is it unproblematic and no contradiction that understanding is an ability, a type of knowledge-how, which requires explanation, which is a type of knowledge-that? The potential conflict that could be assumed here dissolves as soon as it is realized that understanding and knowledge (of explanation) are inseparably intertwined and develop only in conjunction with one another. Based on an extensive discussion of Polanyi's account of the relation between inarticulate and articulate intelligence, I argued that knowledge and understanding necessarily go together because humans cannot make sense of a phenomenon in the world without resorting to the language of the culture in which they were raised and trained. Understanding is an ability and its manifestation, grasping and articulating, is partially tacit and inaccessible to us. This is because we are focally aware of making sense of the phenomenon that we want to understand, we are concentrated on what we observe or measure of the phenomenon, while parallel to our focal attention we subsidiarily make sense of what we perceive, observe of the phenomenon we investigate, by reference to our language. We cannot articulate our understanding, that is, how we actually managed to manifest the ability, to make sense of a phenomenon that we investigated, how we grasped specific relations or constructed generative frameworks. But in this tacit process of understanding a phenomenon we resort to explicitly articulated and non-tacit resources that our language provides and apply these resources to the phenomenon. This is the case not only for scientific understanding, which his achieved by using the sophisticated and formalized language of the respective discipline, but also for non-scientific understanding. Lay people understand the world in terms of the language they grew up with. Although the process of gaining understanding of a phenomenon, of arriving at an adequate interpretation through the manipulation of our language, respects certain context-sensitive criteria that guide the permissible use of a language, these criteria do not prescribe any concrete procedure of how one should gain understanding or at which interpretation one should arrive in the end. In short, knowledge-that is required for manifesting understanding, a type of knowledge-how, and through understanding knowledge-that gets expanded, improved, or revised.

Where are we now? After arguing in chapter three that scientific understanding requires explanation, I argued in this chapter that (scientific) understanding is an ability that is manifested in the process of grasping relation of a phenomenon and articulating these relations as explanations. While my argumentation in chapter three is exclusively targeted at scientific understanding, my analysis in chapter four is broader and addresses understanding in general. Although these two different foci of my investigation do not result in a conflict, as I hopefully showed throughout this chapter, one might still doubt whether the conception of understanding I put forward is actually able to capture understanding gained in science. So far, I worked myself through various arguments and developed my own. And yet, the best and most consistent argument loses its relevance if it cannot be related to what is happening in the world. Hence, it is time to look at science itself and see whether my conception of scientific understanding can withstand scientific practice. In the next chapter, I turn to an episode from biology, the introduction of zebrafish as a model organism, and analyze how scientists gained understanding of the genetic regulation of vertebrate development with the use of zebrafish.

# 5. Scientific understanding of the genetic regulation of vertebrate development and how zebrafish made it possible

So far, I argued that scientific understanding requires explanation and that understanding should be conceived as an ability that is manifested in grasping relations of the phenomenon to be understood and articulating these relations in the form of explanations. While my argumentation has hopefully convinced at least some readers, a central and justified question remains: does my view capture actual cases of understanding in scientific practice? This issue becomes even more pressing in light of the fact that I do not necessarily limit my analysis in the previous two chapters to scientific understanding. Especially in chapter four, I argued that understanding generally should be conceived as an ability. And in chapter three, I addressed several arguments concerning the relation between understanding and explanation without exclusive reference to scientific understanding. Although I do not argue that all kinds of understanding necessarily require explanation, I do claim that scientific understanding does. The basic worry that arises is that, while my arguments might in principle be convincing, they might miss important features or characteristics of scientific understanding and, hence, might not accommodate understanding actually gained by scientists. Even the examples from science that I give in the previous two chapters cannot completely dispel this concern. This is because the function of the examples is to illustrate and substantiate certain philosophical claims, but not to provide important insights into scientific practice. To address the concern that my arguments might not account for understanding that scientists actually achieve in practice or that I might miss important factors or characteristics of understanding, I turn to a concrete and detailed episode from scientific practice in this chapter.

Nowadays it is known that the physiological development of organisms is caused and regulated by genes (amongst other factors). But a genetic understanding of developmental processes is relatively novel and became broadly established only in the 1990s. Understanding developmental and embryological processes as genetically regulated was made possible by the combination of molecular genetics and developmental biology in the late 1960s. Before this rapprochement, scientists working in these two disciplines were interested in and investigating completely different phenomena and did not cooperate. The vision of a better understanding of physiological and embryological processes as well as of genetic functions could only arise due to the specific circumstances in the '60s. From then on, the realization of that vision took 30 years and it relied crucially on one specific model organism: the zebrafish.

In section 5.1, I will briefly present the history of the research around zebrafish in order to understand the genetic regulation of developmental and embryological processes in vertebrates. In doing so, I rely heavily on Robert Meunier's depiction of the development of zebrafish as a model organism. Whereas Meunier is interested in the sense in which model organisms are models, I am using his case study as a basis to trace how the scientists actually gained the understanding of biological phenomena to which they aspired. While the analysis of the research episode will take place in section 5.2, I highly recommend to all readers to not skip section 5.1. I am aware of the potential worry of some readers that they will not understand everything I describe in section 5.1, especially if they do not have a background in biology, or that it would be a waste of time, as I will refer to the most important information again in section 5.2. Although both may be the case, I nevertheless strongly encourage all readers to read both sections in the given order. Having an idea of what happened in the course of this research episode when, in which order and why will make it more easy to then follow the philosophical analysis in section 5.2, and to understand why I emphasize certain details of the research episode in relation to specific philosophical claims. I argue in section 5.2 that the episode from biological research not only supports my arguments from the previous two chapters, that scientific understanding should be conceived of as an ability and requires explanation, but also reveals characteristic features of scientific understanding. In particular, the episode shows that, in order to scientifically understand a specific phenomenon, scientists need to possess relevant pieces of knowledge, research skills and equipment, as well as being situated in an appropriate research infrastructure that ensures functioning communication among scientists and the distribution of resources. Furthermore, the episode also reveals the iterative nature of the manifestation of scientific understanding. That is, scientific understanding does not manifest in a two-step process, consisting of first grasping relations and then articulating explanation. Rather, these two aspects of the manifestation are interwoven and interdependent.

But before turning to these characteristics of scientific understanding, one might wonder why I chose this particular scientific episode.[1] My main motivation

1 In using the term 'episode' instead of 'case study', I follow Hasok Chang in his attempt to address issues in the field of Integrated History and Philosophy of Science. According to Chang, "it is instructive to try seeing the history-philosophy relation as one between the *concrete* and the *abstract*, instead of one between the particular and the general. Abstract ideas are needed for the understanding of *any* concrete episode, so we could not avoid them even if we only

when I started looking for an episode from scientific practice was to find a candidate that is not from physics. Episodes from physics dominate in the philosophical literature on scientific understanding at least so far. All the extensive case studies I am aware of are from physics, and many other authors focus on physics, too, when they present shorter examples in their philosophical texts. Henk de Regt, Kareem Khalifa, and Finnur Dellsén, whose accounts I present in chapter two, all refer mainly to episodes from physics. While other (and often shorter) examples from different disciplines like biology or climate science occasionally appear in the debate on understanding, my impression is that physics still occupies a special status.

This dominance of physics is problematic. If the topic of philosophical analysis is *scientific* understanding, and this topic is approached mainly on the basis of episodes from *physics*, this procedure might lead to biases concerning the nature or acquisition of understanding in various different scientific disciplines. Relying mainly on episodes from physics might result in views or accounts of scientific understanding that suits understanding gained in physics very well, but that, by closer examination, might not accommodate understanding gained in the various branches of biology, climate science, psychology, or the social sciences and so on. I am not denying that understanding gained in various scientific disciplines might not share some fundamental characteristics. After all, I develop an account of *scientific* understanding, understanding gained in science in general. However, in order to identify the fundamental common characteristics of understanding gained in diverse scientific disciplines, philosophers of science should also pay attention to this diversity. And we have the resources to do this. Sub-fields like philosophy of the life sciences, of climate science, and of the social sciences developed in part because of the recognition of the diversity of different scientific disciplines. Philosophers of science interested in scientific understanding should of course also look at physics, but given the attention to physics in the literature on scientific understanding so far, I prioritize increasing the focus on other scientific disciplines. I contribute to this development with the episode from biology I engage with here.

The second reason why I find this scientific episode about zebrafish particularly interesting is the possibility to directly engage with the phenomenon that shall be understood. I will clarify in the course of section 5.1 what exactly I mean by this. In a nutshell, and in contrast to most other episodes or examples from science found

ever had one episode to deal with. […] Any concrete account requires abstract notions in the characterization of the relevant events, characters, circumstances and decisions. If we extract abstract insights from the account of a specific concrete episode that we have produced ourselves, that is not so much a process of *generalization*, as an *articulation* of what we already put into it. To highlight this change of perspective, I prefer to speak of historical "episodes" rather than "cases"." Chang, H. (2012), "Beyond Case-Studies: History as Philosophy." In Schmaltz, T. & Mauskopf, S. (eds.), *Integrating History and Philosophy of Science: Problems and Prospects*, pp. 109–124, Dordrecht, Springer, DOI: 10.1007/978-94-007-1745-9_8, p. 110, original emphasis.

in the philosophical literature on scientific understanding, scientists in the episode about zebrafish actually manipulated *real* instances of the phenomenon they wanted to understand in *real* fish that exhibit this phenomenon. That is, scientists were able to do things that are impossible to do when only theoretical models like mathematical equations or computer simulations are used in research. In such cases, scientists can manipulate the models and make inferences to the phenomena these models represent, but they cannot manipulate the phenomena themselves. This feature of the research on model organisms, the possibility to literally operate on the phenomenon under investigation, really made me interested in how biologists work with zebrafish. So, let us take a look at the episode from science itself and see what happened with and around zebrafish in biology.

## 5.1 How zebrafish became a model organism: the integration of molecular genetics and developmental biology

In the history of zebrafish as a model organism, Meunier identifies three stages that seem to apply to the development of most model organisms. I adopt this partition for my analysis. Meunier characterizes these stages in the following way:

> 1. The choice and introduction of the organism into research [...] and its stabilization in research programmes like neuro-physiology, developmental or cell biology, which are integrative in the sense that they deal with phenomena on many different levels of biological organization and therefore recruit practices from a variety of fields. This stage includes the development of core descriptive and manipulative tools.
>
> 2. The accumulation of large collections of mutant strains and genomic data, and the construction of an infrastructure to maintain and share data and material resources.
>
> 3. The actual use of the model organism to construct models of mechanisms and the generalization of the mechanism by remodelling them in other organisms and constructing abstract mechanism schemata.[2]

2 Meunier, R. (2012), "Stages in the development of a model organism as a platform for mechanistic models in developmental biology: Zebrafish, 1970–2000." *Studies in History and Philosophy of Biological and Biomedical Sciences*, 43, pp. 522–531, DOI: 10.1016/j.shpsc.2011.11.013, p. 523. Over the past decades, a mechanistic explanation paradigm has been established in biology. Biological phenomena are explained in terms of mechanisms that specify, for example, underlying parts of the phenomenon, their organization, or their interaction. The case discussed here is an instance of this paradigm. I accept this paradigm and will not analyze or criticize it. For more information concerning the mechanistic explanation paradigm

### 5.1.1 Choosing, introducing, and stabilizing zebrafish in research

Meunier discusses all the three stages in more detail. To begin with, it was probably an influential factor that zebrafish had already been introduced as a research organism (but not a model organism) in the 1930s at the latest.[3] This distinction between research and model organism matters, because these two different types of organisms are used to study and understand different phenomena. Rachel Ankeny & Sabina Leonelli present the following differentiation between research or experimental organisms and model organisms:

> In short, although both experimental and model organisms are models in the sense of being representative of a larger class of organisms, they are distinct types of models because of the fundamental difference in the breadth of their representational scope and, most importantly, their intended representational target. Experimental organisms tend to be models for particular phenomena, while *model organisms are models for organisms as wholes*, used not just to explore specific phenomena, but *aimed at developing an integrative understanding of intact organisms* in terms of their genetics, development, and physiology, and in the longer run of evolution and ecology, among other processes.[4]

So, according to Ankeny & Leonelli, the zebrafish was introduced as a model organism because scientists wanted to understand intact organisms. If this were not the

---

in biology, see Machamer, P., Darden, L. & Craver, C. F. (2000), "Thinking about Mechanisms." *Philosophy of Science*, 67 (1), pp. 1–25, DOI: 10.1086/392759; or Bechtel, W. & Abrahamsen, A. (2005), "Explanation: a mechanist alternative." *Studies in History and Philosophy of Biological and Biomedical Sciences*, 36, pp. 421–441, DOI: 10.1016/j.shpsc.2005.03.010; or Darden, L., (2008), "Thinking Again about Biological Mechanisms." *Philosophy of Science*, 75 (5), pp. 958–969, DOI: 10.1086/594538; among others.

3 See Meunier (2012), p. 524. For more information, see Creaser, C. W. (1934), "The technic of handling the zebra fish (Brachydanio rerio) for the production of eggs which are favorable for embryological research and are available at any specified time throughout the year." *Copeia*, 4, pp. 159–161, DOI: 10.2307/1435845. For an overview on the use of zebrafish in science before its establishment as a model organism, see Laale, H. W. (1977), "The biology and use of zebrafish, Brachydanio rerio in fisheries research. A literature review." *Journal of Fish Biology*, 10, pp. 121–173, DOI: 10.1111/j.1095-8649.1977.tb04049.x.

4 Ankeny, R. A., & Leonelli, S. (2011), "What's so special about model organisms?" *Studies in History and Philosophy of Science*, 42 (2), pp. 313–323, DOI: 10.1016/j.shpsa.2010.11.039, p. 319, my emphasis. For more detailed information about the difference between experimental and model organisms see ibid. Although the authors put quite a lot of emphasize on understanding, they do not analyze this concept further. Still, their statement serves as supporting evidence that there is something epistemically interesting and important about the understanding of intact organisms via model organisms and that this subject should be analyzed in more detail.

case, zebrafish would not be appropriately viewed as a model organism. Let's keep this in mind and see what the scientists involved in the research on zebrafish actually wanted to achieve with this organism.

The person said to have initiated the development of zebrafish as a model organism was George Streisinger (1927–1984), a phage geneticist. Together with some colleagues, he was at the heart of phage genetics and involved in the emergence of the new field of molecular biology in the 1960s. Streisinger started to work with zebrafish with the goal to "establish zebrafish as a system that would allow him to relate the knowledge of molecular genetics that he had helped to establish to complex organismic properties."[5] Why did he choose zebrafish, and not some other experimental organism? Looking at the history of different model organisms, a sound list of appropriate features that facilitate the intended research can be identified: small size, short generation time, large amounts of eggs every week throughout the year, rapid development outside of the mother, and robustness to environmental influences, among others. These are instrumental traits, traits that make it easier for scientists to conduct their studies, some of which are shared by zebrafish and other organisms. The crucial feature that makes zebrafish especially suitable for developmental studies is that, during the first stages of development, the embryos and larvae are transparent. For that reason, it is relatively easy to study organogenesis, the phase of embryonic development during which the internal organs of an organism are formed from the three germ layers, with a simple dissection microscope. An additional important factor for Streisinger in choosing zebrafish was that fish seemed to be a good compromise. Since Streisinger wanted to conduct research on vertebrates, he needed a model organism closer to larger vertebrates but that is still small enough and reproduces quickly and in sufficiently large numbers to apply genetic strategies. Zebrafish is gigantic in comparison to, for example, fruit flies, but small for a vertebrate. But why zebrafish, since many fish share the important features of external fertilization and development? What made Streisinger ultimately decide on zebrafish and not another fish species is unknown. Maybe it was just a matter of chance and maybe other fish would have served the purpose just as well. The more important and crucial question is: why did Streisinger move from phage genetics to multicellular organisms in the first place?[6]

---

5 Meunier (2012), p. 524. For more biographical information about Streisinger see Stahl, F. W. (1995), "George Streisinger—December 27, 1927—September 5, 1984." *Biographical Memoirs. National Academy of Sciences (U.S.)*, 68, pp. 353–361; and Endersby, J. (2007), *A guinea pig's history of biology*. London, William Heinemann Ltd, chapter 11. For more details about the history of molecular biology, see Cairns, J., Stent, G. S. & Watson, J. D. (eds.) (1966), *Phage and the origins of molecular biology*. Cold Spring Harbor (NY), Cold Spring Harbor Laboratory Press.

6 See Meunier (2012), p. 524.

The reason for Streisinger's new interest was the view of bacteria and phage geneticists in the 1960s of having arrived at a dead end. No one expected any new groundbreaking findings in the discipline of molecular biology after the structure of DNA and the genetic code had been established. Molecular biology has turned into a normal science in which details are sorted out under a given paradigm.[7] Many bacteria and phage geneticists saw only one option, which Sidney Brenner characterized as "the extension of research to other fields of biology, notably development and the nervous system."[8] Therefore, many molecular biologists, Streisinger among them, started to work on more complex organisms, like mice, *Drosophila* or *C. elegans*, in order to extend the scope of their discipline. There was great optimism among the contemporaries about establishing this new research program of developmental or neuro-genetics.[9] For example, Brenner stated that "in principle, it should be possible to *dissect* the genetic specification of a nervous system in much the same way as was done for biosynthetic pathways in bacteria or for bacteriophage assembly."[10] Seymour Benzer described his vision in even more detail:

> Once assembled, the functioning nervous system embodies a complex of interacting electrical and biochemical events to generate behaviour. The fine structure and interlacing of even the simplest nervous systems are such that to *dissect* them requires a very fine scalpel indeed. *Gene mutation* can provide such a *microsurgical tool*; with it one might hope to analyse the system in a manner *analogous* to the one which has proven so successful in unravelling *biochemical pathways* and control *mechanisms at the molecular level*. [...] Among a collection of such non-phototactic mutants, one might expect to find defects affecting the *various elements of the system*. [...] This search for defects in non-phototactic mutants describes the outline of *a research program to attack the mechanisms underlying behaviour by genetic methods*.[11]

7 The situation in late 19th century physics, before the emergence of quantum and relativity theory, was comparable to the situation in molecular genetics in the 1960s. Late 19th century physicists also thought that they knew everything about the physical world that there is to know and that the future task of physicist will be limited to the more detailed specification of natural constants. As things turned out, this expectation was wrong.

8 Brenner, S. (1998), "Letter to Perutz." In Wood, W. B. (ed.), *The nematode Caenorhabditis elegans*, pp. x-xi, Cold Spring Harbor (NY), Cold Spring Harbor Laboratory Press, p. x.

9 See Meunier (2012), pp. 524f. I am adopting the term research program from Meunier as well as from some molecular biologists, but I am not advocating a Lakatosian theory of science. Occasionally, I refer to the new emerging scientific research as a discipline. It does not make a difference for my point about scientific understanding whether the integration of molecular genetics with developmental biology is labelled research program, discipline, or otherwise.

10 Brenner, S. (1974), "The genetics of Caenorhabditis elegans." *Genetics*, 77 (1), pp. 71–94, DOI: 10.1093/genetics/77.1.71, p. 72, my emphasis.

11 Benzer, S. (1968), "Genes and behavior." *Engineering and Science*, 32, pp. 50–52, my emphasis.

Meunier emphasizes that the important methodological metaphors used by the molecular geneticists were 'dissection' and 'surgery', while the metaphors 'pathway', 'circuit' and 'mechanism' designated the phenomena to be investigated by the methodology. These metaphors represent the vision that molecular geneticists had of their new research program. And most importantly, both Brenner and Benzer already pointed towards the goal of dissecting not only molecular mechanisms that underly organismic phenomena, but also processes on a higher physiological level, especially related to the nervous system.[12]

However, the realization of that vision was not as easy as molecular geneticists had hoped. The attempts to apply molecular knowledge to more complex organisms like mice, *C. elegans* or zebrafish were at first disenchanting. The successful establishment of the new research program of developmental and neuro-genetics, together with the establishment of model organisms like zebrafish within this program, was due in the end to an integration of molecular genetics with classical embryology and neuro-physiology. Importantly, embryologists and classical geneticists working on development also broadened their view and started to investigate molecular processes *independently* of molecular geneticists. Already since the 1950s, the term 'developmental biology' was used to refer to "the broadening of interests and the integration of different biological disciplines, in particular genetics, biochemistry, classical experimental embryology and molecular biology."[13] When molecular geneticists became interested in higher organisms, they relied heavily on the traditional practices, questions, and expertise of embryologists and physiologists, since the concepts and orientation of molecular biology had changed. Conventionally, molecular biology was interested in the structural and informational basis of replication and the synthesis of cellular molecules, for which specific practices were required. With the emergence of the new research program of developmental and neuro-genetics, the descriptive level moved from molecules to cells. Cell activities were the new explananda at which research on molecular activity was aiming. Instead of merely asking how DNA replicates, how proteins are synthesized and how they interact in metabolic reactions, molecular geneticists aimed to understand these activities in the context of complex phenotypes. As yet, these chemical activities were only related to phenotypes of bacteria and phages. These phenotypes are often defined as the growth of a bacterial colony under specific

---

12 See Meunier (2012), p. 525.

13 Fantini, B. (2000), "Molecularizing embryology: Alberto Monroy and the origins of developmental biology in Italy." *The International Journal of Developmental Biology*, 44 (6), pp. 537–553, p. 548. Alfred Kühn, Joseph Needham and Conrad Waddington, or Jean Brachet were among those early developmental biologists who were interested in molecular processes before the fields of molecular genetics and developmental biology merged.

circumstances, and the relation between these phenotypes and chemical activities is relatively straightforward. In the case of multicellular organisms, however, molecular geneticists had to handle phenotypes at many levels of organization, from cellular, tissue and organ properties up to the behavior of the organism as a whole. Molecular geneticists needed the expertise from embryologists and neurophysiologists in order to manage the new phenotypes with which they had not previously been concerned.[14]

The establishment of zebrafish as a model organism is a case of this integration between molecular genetics and developmental biology. Streisinger's lab at the University of Oregon first developed ways to reliably maintain the zebrafish colonies. After this was achieved, the main goal of the lab was to establish tools for the genetic analysis of zebrafish. The first great success was the development of a technique that enabled the scientists to use artificial parthenogenesis to produce homozygous diploid animals. With this technique, clonal strains for later mutational analysis could be generated. These clonal strains were free of lethal mutations. The next tasks were to introduce mutations in the zebrafish strains and to devise mapping strategies for the zebrafish mutations. The main reason for Streisinger to develop all these new techniques was the "dissection of neuronal development by the use of mutant strains."[15] This objective was realized in the mid-80s, in form of $\gamma$-ray mutagenesis experiments, which was used to analyze a neuronal necrosis mutant. The results were published in 1988. Importantly, this analysis was a cooperation between Streisinger's lab at the Institute of Molecular Biology and the Institute of Neuroscience, both located at the University of Oregon. Monte Westerfield and Charles Kimmel were the members from the Institute of Neuroscience who participated in the $\gamma$-ray mutagenesis experiment together with the molecular geneticists. Notably, Kimmel has been trained as a developmental biologist. The molecular geneticists working in Streisinger's team did not reach their goal of the dissection of neuronal development on their own.[16] Meunier very nicely summarizes the integrative character of this research endeavor:

> Whereas Streisinger's lab brought in the expertise needed for the generation and genetic characterization of the mutation (segregation analysis, karyotype analysis), the description of the phenotypic effects of the mutation was based on

---

14 See Meunier (2012), p. 525.

15 Streisinger, G., Walker, C., Dower, N., Knauber, D. & Singer, F. (1981), "Production of clones of homozygous diploid zebra fish (Brachydanio rerio)." *Nature*, 291 (5813), pp. 293–296, DOI: 10.1038/291293a0, p. 293.

16 See Meunier (2012), pp. 525f. For more details concerning this series of experiments, see Grunwald, D. J., Kimmel, C. B., Westerfield, M., Walker, C. & Streisinger, G. (1988), "A neural degeneration mutation that spares primary neurons in the zebrafish." *Developmental Biology*, 126 (1), pp. 115–128, DOI: 10.1016/0012-1606(88)90245-x.

> the knowledge that Westerfield and especially Kimmel had accumulated over the preceding 15 years. They provided the descriptive knowledge and methods to behaviourally and physiologically characterize the specific effect of the mutation (fixing the embryos, measuring neuromuscular activity, staining sections, conducting behavioural response tests). The articulation of the main result — that the mutation selectively affected the central nervous system by causing necrosis in most cells, but sparing a particular class of neurons, namely primary neurons — required detailed knowledge of the neuro-anatomy of the fish, which Kimmel and Westerfield did [...] posses.[17]

Molecular geneticists, who moved to multicellular organisms, needed the knowledge and practices from classical embryology and physiology, as without the descriptive work and knowledge provided by these disciplines, mutation-based studies would not have been possible. The reason is that descriptive devices like cellular fate maps, neural wiring diagrams or staging series[18] define the normal or wild-type organism, which refers to a non-manipulated member of the strain from which the manipulated members are also derived. Thus, the wild-type is a standardized strain. Without these sources, scientists would not have a contrastive foil that makes the mutation actually visible through comparison. Kimmel and his colleagues provided a first fate map for zebrafish in 1990 and published a staging series for zebrafish in 1995.[19] However, not only did the molecular geneticists rely on the descriptive practice of embryologists and physiologists in order to study the function of genes, but also embryologists and physiologists took on the methodology of mutational analysis in their research. For example, Kimmel and Westerfield continued to use mutational analysis after the publication of the neural degeneration mutant in 1988. Mutational analysis enabled explanations in terms of molecular genetics, which opened up new ways of research and were helpful in identifying structures, processes and functions on higher levels of organization. Among other things, the analysis of differential effects of a mutation made a more fine-grained classification of cell types

17 Meunier (2012), p. 526.

18 Cellular fate maps are representations that trace the history of each cell in development, neural wiring diagrams are descriptions of a nervous system and staging series define steps in the continuous process of development in embryos. For more details concerning descriptive models and descriptive devices in biological research, see Ankeny, R. A. (2000), "Fashioning descriptive models in biology: of worms and wiring diagrams." *Philosophy of Science*, 67, pp. 260–272, DOI: 10.1086/392824.

19 See Kimmel, C. B., Warga, R. M. & Schilling, T. F. (1990), "Origin and organization of the zebrafish fate map." *Development*, 108 (4), pp. 581–594, DOI: 10.1242/dev.108.4.581; and Kimmel, C. B., Ballard, W. W., Kimmel, S. R., Ullmann, B. & Schilling, T. F. (1995), "Stages of embryonic development of the zebrafish." *Developmental Dynamics: An Official Publication of the American Association of Anatomists*, 203 (3), pp. 253–310, DOI: 10.1002/aja.1002030302.

possible. Furthermore, mutation based analysis have become mainstream in developmental biology after the successful molecularization of *Drosophila* and *C. elegans* embryology. Therefore, it was probably easier to publish mutation-based studies than classical physiological studies by the 1990s.

At this point, the first stage of the development of zebrafish as a model organism, the choice and introduction of the organism into research and its stabilization in research programs was completed. Zebrafish was a model system for mutational analysis of development and physiological processes. For the next stage, the accumulation of more data, the material resources and the necessary infrastructure to maintain data and resources had to be developed.[20]

## 5.1.2 Building and establishing a research infrastructure

This happened in the mid-1990s through two coordinated large-scale mutagenesis screens, called *The Big Screen* in the zebrafish community. However, *The Big Screen* required some more preparative work. At the core of the integration of molecular genetics with embryology, right from the start, was the technique of mutational dissection. Mutational dissection enabled scientists to identify single genes that participate in the development of certain traits and the molecular characterization of these genes. Yet mutational dissection was crucially limited in the sense that scientists aimed at molecular explanations of development that include *interactions* between molecules that result in certain cellular behavior. These explanations only became available in late 1970s and Christiane Nüsslein-Volhard and Eric Wieschaus made a huge contribution to this breakthrough. They conducted a systematic search for mutations that affect embryonic patterning in *Drosophila* by a large-scale screen for mutants. Nüsslein-Volhard and Wieschaus selected only those mutants that showed an effect in embryonic patterning in order to detect the relevant genes for that process. Both researchers did not rely on any preceding ideas about which genes might be involved or how they influence embryonic patterning. This is called a saturation screen for specific phenomena, embryonic patterning in this case, and resulted in collections of different mutant strains. The identification of genes that affect the phenomenon when mutated enabled the scientists to hypothesize interactions among genes and molecular level explanations.[21] The hypothesized interactions with regard to regulatory pathways and mechanistic explanations on a molecular level were investigated in subsequent studies by using different collec-

20 See Meunier (2012), p. 526.

21 For more information concerning the experiments conducted by Nüsslein-Volhard and Wieschaus, see Nüsslein-Volhard, C. & Wieschaus, E. (1980), "Mutations affecting segment number and polarity in Drosophila." *Nature*, 287 (5785), pp. 795–801, DOI: 10.1038/287795a0.

tions of mutant strains.[22] The first project from Nüsslein-Volhard and Wieschaus achieved explanation in terms of molecular regulatory mechanisms at a depth that had not previously been achieved.[23]

Why was this study on *Drosophila* conducted by Nüsslein-Volhard and Wieschaus important for the research on zebrafish and how was it connected to *The Big Screen*? By the early 1990s, the required genetic and descriptive tools for the envisioned research had been implemented, but only a small number of zebrafish mutants was available. It was Nüsslein-Volhard who had the idea that a project very similar to the *Drosophila* screens could be used to obtain more zebrafish mutants when she read Streisinger's 1981 paper. Nüsslein-Volhard subsequently started to develop the infrastructure for a large-scale mutagenesis screen in zebrafish at the beginning of the 1990s. This included, among other things, the construction of new aquaria systems. Nüsslein-Volhard and her colleagues invested an extensive amount of creativity in this research infrastructure and published their results of a first pilot screen together with the specifications of the screen in 1994.[24] Thereby, *The Big Screen* was initiated in Tübingen in 1993 under the supervision of Nüsslein-Volhard. A very similar project was launched at the Massachusetts General Hospital in Boston under the direction of Wolfgang Driever, a former student of Nüsslein-Volhard who had worked with her on *Drosophila*. These two coordinated large-scale mutagenesis screens in Tübingen and Boston are the research projects labelled *The Big Screen*.[25]

In order to detect mutants, scientists involved in *The Big Screen* first observed standard anatomical features of the fish under a dissection microscope. The standard anatomical features were defined with a descriptive device, a check list that showed a simple anatomic map of zebrafish. The observed anatomical features of mutants were compared to wild-type animals that were raised under the exact same conditions as the mutants. Various stains were used in the subsequent steps to determine more fine-grained differences among similar phenotypes. The guiding heuristic assumption was that genes, which produce related phenotypes when mutated, might react with each other under normal conditions. Therefore, mutations with similar phenotypes or those with an effect on the same structures were grouped together, resulting in a large number of zebrafish mutant strains. The results from both screens were published together in 1996.[26] The crucial im-

22 For more information about follow up studies, see Driever, W. & Nüsslein-Volhard, C. (1989), "The bicoid protein is a positive regulator of hunchback transcription in the early Drosophila embryo." *Nature*, 337 (6203), pp. 138–143, DOI: 10.1038/337138a0.

23 See Meunier (2012), p. 527.

24 For more details, see Mullins, M. C., Hammerschmidt, M., Haffter, P. & Nüsslein-Volhard, C. (1994), "Large-scale mutagenesis in the zebrafish: in search of genes controlling development in a vertebrate." *Current Biology*, 4 (3), pp. 189–202, DOI: 10.1016/s0960-9822(00)00048-8.

25 See Meunier (2012), p. 527.

26 See ibid. p. 527.

portance of the availability of the mutant strains was expressed, among others, by Philip Ingham, who said that "the identification of so many mutations affecting zebrafish embryogenesis represents a quantum leap in our capacity to unravel the mechanisms underlying vertebrate development."[27] From then on, scientists could simply choose those mutations that affect the developmental process of interest. As a result, many postdocs who worked on the two mutagenesis screen projects took a set of related mutations and founded new labs, where they used the zebrafish mutants to investigate the mechanisms in which certain genes interact.[28] Hence, by the late 1990s the second stage in the development of zebrafish as a model organism, the accumulation of large collections of mutant strains and the construction of a research infrastructure, had been completed.

### 5.1.3 Using zebrafish as a model organism

In the third and final stage, the model organism was used, finally, to construct models of mechanisms. This was possible only because the manipulative and descriptive tools had been developed at the first stage, and the large-scale mutagenesis screens had been performed at the second stage to identify and provide different mutant strains. To show how the third step was realized, Meunier presents the research on one of the mutants that was identified in the screen, *one-eyed pinhead* (*oep*).[29]

The research groups in Tübingen and Boston identified different alleles of the *oep* gene. Like most other mutants, the *oep* mutant was assigned to more than one class of mutant phenotypes due to the different processes or structures affected by *oep* at different stages in development. The next success following the identification of *oep* was achieved by the Driever's group in Boston. They could, on the one hand, specify how *oep* affects the formation of the three primary germ layers during gastrulation, a process taking place in the early embryogenesis. On the other hand, by creating a double mutant, they showed a genetic interaction between *oep* and the *no tail* (*ntl*)

---

27 Ingham, P. W. (1997), "Zebrafish genetics and its implications for understanding vertebrate development." *Human Molecular Genetics*, 6 (10), pp. 1755–1760, DOI: 10.1093/hmg/6.10.1755, p. 1759.

28 See Meunier (2012), p. 527.

29 Following the convention in the field, gene symbols are lower case and italicized, while protein symbols are the same as the corresponding gene symbols, but the first letter is uppercase and the protein symbols are non-italic, see (https://wiki.zfin.org/display/general/ZFIN+Zebrafish+Nomenclature+ConventionsZFINZebrafishNomenclatureConventions-2; last accessed April 12th, 2022).

gene.[30] Then, other researchers in New York achieved the molecular isolation of *oep* using positional cloning.[31] That was the first time that a Zebrafish gene was cloned by this approach.[32]

The cloned *oep* mutant embryos were lacking Oep activity, which leads to defective germ layer formation, organizer development, and positioning of the anterior-posterior axis that results in a cyclopic phenotype without endoderm, prechordal plate, and ventral neuroectoderm.[33] Meunier describes the importance of the molecular isolation of *oep* as follows:

> This allowed comparing the sequence of the Oep protein to other known proteins, which suggested that it had a signalling and a membrane binding sequence. At the same time it allowed applying many molecular strategies, like injection of the mRNA for rescue or overexpression, or fusion mRNA's coding for markers, detectable by immunostaining or otherwise, as well as *in situ* hybridization to observe expression patterns. These techniques were immediately used to localize the protein on the cellular level and the expression of the gene in the embryo.[34]

In the course of subsequent research, it was discovered that Oep is an essential component of the Nodal signaling pathway. The signaling molecule Nodal plays an important role in early embryonic patterning and has been discovered for the first time in mice. In zebrafish, two orthologs[35] of *nodal*, *cyclops (cyc)* and *squint (sqt)*, were found. The products of *cyc* and *sqt* are collectively called Nodal signal. The crucial observation in these experiments was that the phenotype of the double mutant for the *cyc* and *sqt* genes and the *oep* mutant phenotype are very similar. This observed similarity, together with the fact that embryonic processes associated with Nodal signaling are affected by the *oep* mutation, and with the knowledge that Oep is membrane-associated while acting cell-autonomously[36], resulted in the hypothesis

---

30 For more information, see Schier, A. F., Neuhauss, S. C., Helde, K. A., Talbot, W. S. & Driever, W. (1997), "The one-eyed pinhead gene functions in mesoderm and endoderm formation in zebrafish and interacts with no tail." *Development*, 124 (2), pp. 327–342, DOI: 10.1242/dev.124.2.327.

31 For more detailed information, see Zhang, J., Talbot, W. S. & Schier, A. F. (1998), "Positional cloning identifies zebrafish one-eyed pinhead as a permissive EGF-related ligand required during gastrulation." *Cell*, 92 (2), pp. 241–251, DOI: 10.1016/S0092-8674(00)80918-6.

32 See Meunier (2012), p. 528.

33 See Gritsman, K., Zhang, J., Cheng, S., Heckscher, E., Talbot, W. S. & Schier, A. F. (1999), "The EGF-CFC protein one-eyed pinhead is essential for nodal signaling." *Cell*, 97 (1), pp. 121–132, DOI: 10.1016/s0092-8674(00)80720-5, p. 121.

34 Meunier (2012), p. 528.

35 Genes in different species, which originated from a single gene from the last common ancestor of these species by vertical descent, are termed orthologs or orthologous genes.

36 This knowledge had been established by other studies, see for example Schier et al. (1997).

that "Oep is required for cells to receive Nodal signals."[37] In order to arrive at this hypothesis, it was necessary to carefully compare and relate different mutants from the large collection available. However, to understand how exactly Oep affects Nodal signaling, the scientists wanted to figure out whether Oep is necessary for Nodal signaling, or whether it merely has an amplifying function, and where exactly it is located in the biochemical pathway. Two experiments were performed to obtain insights about this process.[38]

First, to test whether Oep is necessary or merely an amplifier for Nodal signaling, an overexpression of the Nodal signal was induced through the injection of *cyc* and *sqt* mRNA. If Oep is an amplifier, this overexpression would lead to development of the *oep* mutant such that its phenotype would be closer to the wild type, or even lead to dorsalization.[39] In other words, the idea was that the injected *cyc* and *sqt* mRNAs would replace the function of *oep*, which is absent in the *oep* mutants, and, therefore, the defects in the *oep* mutant embryos due to the absence of *oep* would be corrected and normal, wild-type phenotypes would develop. However, no effect was observed in *oep* mutants, which led to the conclusion that, during embryogenesis, Oep is indeed essential for Nodal signaling and not merely an amplifier. Sqt/Cyc do not replace the function of Oep. For the second experiment, the scientists already had evidence for the transmission of Nodal signaling in the cell by a pathway that involves the ActRIB receptor and the Smad2 transcription factor. The goal was to determine whether this evidence was correct, whether Oep is indeed essential for the response to these two factors. Therefore, mRNAs of the genes that code for these factors (already activated versions of ActRIB and Smad2) were injected in *oep* mutants. The activation of this pathway by Oep was simulated by the injection of the activated factors. In this case, the mutant phenotypes became more similar to the wild type. The hypothesis that Nodal signals are transmitted by this pathway was confirmed and the experiment showed that Oep acts upstream of these components. Hence, "Oep [is identified] as a novel and specific component of the Nodal signalling pathway"[40] and Oep was localized as an extracellular co-factor, which is necessary for the Nodal signal to activate the downstream elements in the pathway. In follow-up studies that used further mutants and reagents, further elements of the pathway

---

37 Gritsman et al. (1999), p. 125.

38 See Meunier (2012), p. 529. The whole study consisted of more experiments, which are all presented in Gritsman et. al. (1999). However, the two experiments presented here were the most crucial ones.

39 Dorsalization refers to the formation of dorsal cell types, one class of primary sensory neurons in the lamprey spinal cord, and the organization of tissues along the dorsoventral (from back to belly) axis.

40 Gritsman et. al. (1999), p. 128.

were added.[41] However, to achieve the bigger goal of explaining organizer function in early embryonic patterning, the investigation on the Nodal signaling pathway was only one step. To relate such pathways to cellular phenotypes and to explain broader developmental or physiological processes on that basis, the same material models, i.e. the same sets of mutants and reagents, could be used in varying combinations with other components, like appropriate cameras, for the respective level of biological organization.[42]

Importantly, in the two experiments on the *oep* mutants situations were observed in which an abnormal phenotype is a result of the absence (loss of function) of the respective genes. On the basis of that observation, scientists established a causal relation between a gene and an aspect of the normal phenotype. But the aim of the experiments was to establish causal relations not between genes and partial phenotypes, but instead among genes. The identified causal relations between genes and phenotypes have an instrumental purpose, because the causal relations among genes can be inferred from the causal relations between genes and phenotypes.[43] For instance, Meunier reconstructs the inference made in the second experiment on *oep* mentioned above:

> "Oep, and everything that comes downstream in the causal chain, cause an aspect of the phenotype (in the precise sense that the absence of Oep and therefore of the activity of downstream elements results in an aberration). Whatever is downstream of Oep in the causal chain is not active if Oep is absent. If the normal phenotype is present if ActRIB and Smad2 are present (enforced, independently of Oep), in the absence of Oep, then they should act downstream of Oep in the causal pathway."[44]

The causal relations between genes and partial phenotypes provide the basis for abductive inferences about the causal interactions among proteins produced by certain genes. These interactions are the regulatory events that allow for cell differentiation. Achieving the goal of identifying and establishing causal relations among genes, therefore, requires counterfactual reasoning, abductive inference and the paradigm

41 For more information, see Bisgrove, B. W., Essner, J. J. & Yost, H. J. (1999), "Regulation of midline development by antagonism of lefty and nodal signaling." *Development*, 126 (14), pp. 3253–3262, DOI: 10.1242/dev.126.14.3253.

42 See Meunier (2012), p. 529. Such a study that involved *oep* mutants and other components of the zebrafish Nodal signaling model has been conducted, for example, by de Campos-Baptista, M. I., Holtzman, N. G., Yelon, D. & Schier, A. F. (2008), "Nodal signaling promotes the speed and directional movement of cardiomyocytes in zebrafish." *Developmental Dynamics: An Official Publication of the American Association of Anatomists*, 237 (12), pp. 3624–3633, DOI: 10.1002/dvdy.21777.

43 See Meunier (2012), p. 529.

44 Ibid. p. 529.

of differential gene expression.[45] In case of the *oep* mutant, a model of the Nodal signaling was successfully constructed. With that, the third and final stage of the development of zebrafish as a model organism was reached. If, in a next step, the knowledge gained by these experiments is to be generalized, such material models of mechanisms have to be instantiated in other species, which will require analogical reasoning. When the mechanism is successfully instantiated in different material models, it still demands inductive inference to arrive at a generalization of the respective molecular mechanism.[46]

### 5.1.4 Explaining physiological phenomena through molecular regulation

Let me summarize this episode from scientific research as it is presented by Meunier. Around 1970 a new research program developed as a combination of developmental biology and molecular genetics with the aim of constructing, through mutational dissection of molecular pathways, models of the genetic regulation of processes in the development, physiology, and behavior of multicellular organisms. Zebrafish was introduced as a new model organism that served this purpose, as illustrated by the example of *oep* mutants to model the Nodal signaling pathway. The new research program included genetic and physiological techniques and the skills to employ them, descriptive devices and nomenclature, collections of mutant strains and sequence data, and the infrastructure to share these resources. In order to determine causal relations among genes, a whole arrangement of mutants, mRNAs and other reagents, and instruments such as microscopes was necessary. The vision of Benzer, Brenner and Streisinger to achieve explanations of physiological phenomena in terms of molecular regulation gained by mutational analysis was fulfilled. The new research community developed manipulative and descriptive tools, generated mutants in various mutagenesis experiments and shared these mutants as well as the available information.[47] Meunier himself concludes:

> Material models of mechanisms are different from the mechanisms themselves [that] occur in the fish, in that the former consist of various objects, animals and

45 This paradigm "implies that gene expression is regulated through gene activity in complex regulatory loops." Ibid. p. 523.

46 See ibid. p. 529. Remember that this is the claim Meunier argues for in his paper. He uses the case study of the development of zebrafish as a model organism to argue that organisms are model organisms in virtue of their use in the construction of models of particular mechanisms, and not in virtue of being models for a higher class of organisms. Again, since my analysis focuses on what scientific understanding is and how it is achieved, I am leaving the question in which sense organisms are model organisms aside and will not discuss Meunier's argument.

47 See ibid. pp. 529f.

> others, that by virtue of their combination and arrangement, carve out the mechanism from the whole of causal interactions taking place in a fish and thus represent them. *But, in contrast to other material models, like plastic ball and stick models of molecules for instance, the fact that these models are built using the organisms that actually exhibit the mechanisms modelled enables researchers to literally operate the mechanism in order to manipulate the developmental process and thereby understand it better.* [...] In this way they generate knowledge about the entities that make up the mechanism (e.g. their activities, their position relative to each other, quantitative characteristics etc.). New entities or activities involved in the mechanism can be added to the model by adding new mutant strains or reagents. *A generative material model as it is described here thus establishes new knowledge through stepwise combination of interventions in the system modelled.* In describing the mechanism as represented in the material model, theoretical models are constructed through text or diagrams.[48]

This quote clarifies two characteristics of model organisms that I mentioned at the beginning of this chapter. First, I claimed that model organisms enable scientists to directly engage with the phenomenon that shall be understood. Scientists do that through manipulating genetic mechanisms that underling embryonic development of real vertebrates. Furthermore, at the beginning of section 5.1.1, I referred to a characterization of model organism proposed by Ankeny & Leonelli. According to them, model organisms are not models for particular phenomena, but rather for organisms as wholes that serve the aim of gaining an integrative understanding of intact organisms. This is exactly what zebrafish was used for in the case of the research on *oep*. Ultimately, biologists were not interested in particular phenomena, such as the effects that the injection of certain mRNA has on some mutant. Instead, researchers wanted to understand the regulatory genetic mechanism underlying developmental, physiological, and behavioral processes *in normal organisms*. They aimed precisely at the integrative understanding of intact organisms that Ankeny & Leonelli demand for model organisms in contrast to experimental organisms. Hence, Meunier's characterization of model organism is in line with the view of Ankeny & Leonelli and it is plausible to regard zebrafish as a new model organism, and not merely an experimental organism.

So much for the historical part. Again, the research question Meunier wants to answer with this episode from scientific practice concerns how model organisms are models or, to put it differently, what 'model' in the term 'model organism' means. While I do not further discuss the concept of model organisms, I use this episode to address a different question: how did the biologists involved in the establishment of the new research program around zebrafish understand the phenomena they wanted to understand and explain? I answer this question in the next section.

48 Ibid. p. 530, my emphasis.

## 5.2 How is this a case of scientific understanding?

What can this episode from scientific research reveal about scientific understanding and the way scientists achieve it? My analysis of the episode around zebrafish serves two goals. First, it serves to examine whether my claims from the two previous chapters, that understanding requires explanation and that understanding is an ability that manifests in the process of grasping relations of a phenomenon and articulating these relations in the form of explanations, are supported or rebutted by the episode from biology. Second, the episode helps to identify characteristics of understanding achieved in science by scientists *qua* scientists that did not become apparent in the preceding chapters, since they did not address scientific understanding specifically. Let's look at the three stages of the establishment of zebrafish as a model organism that Meunier distinguishes and their respective contribution to the possibility of scientific understanding of the genetic regulation of vertebrate development.

### 5.2.1 Gaining necessary knowledge, research skills and tools

The empirical phenomenon that scientists wanted to understand in this case is the genetic regulation of embryonic development of complex organisms like vertebrates. Molecular biologists viewed this phenomenon as a developmental mechanism. The relations that are grasped in this case are causal relations. The genetic activities that take place within a cell cause the development of a specific phenotype of a differentiated cell, and various cell behaviors cause physiological phenotypes.[49] In order to grasp, that is, to get epistemic access to, the causal relations involved in the developmental mechanism, scientists first needed to have the knowledge and skills required for this phenomenon. Recall that I do not see a conceptual difference between the terms 'ability' and 'skill', only a terminological one. However, for the sake of clarity, I will use the term 'ability' to refer to understanding, and the term 'skill' when I refer to any other type of knowing-how that is involved in or serves the goal of understanding.

The availability of the necessary knowledge and skills was only possible because of the integration of molecular genetics with classical embryology and neurophysiology. Molecular geneticists in the 1960s, like Streisinger, Benzer and Brenner, wanted to understand the development of complex organisms, but they were not able to achieve this understanding with the knowledge and skills from their

49 I stay agnostic at this point whether the relations in this case should be seen as causal, mechanistic, or functional relations, since this differentiation does not affect the acquisition of scientific understanding, which requires the grasp of any type of relation. For the sake of convenience, I follow Meunier in taking the grasped relations in this case of scientific research to be causal.

field alone. Molecular geneticists needed the knowledge from embryologists and neurophysiologists as well as the skills to apply research techniques from these disciplines. On the other hand, embryologists and neurophysiologists also had the goal of understanding molecular processes underlying development, yet were similarly not in a position to achieve this goal merely with their own knowledge and skills. Knowledge and skills from both disciplines had to be merged.

This process can be seen clearly in the research conducted during the 1980s by Streisinger's lab. Streisinger's knowledge and skills, which he acquired through his work on phage, were sufficient to find ways of reliably maintaining the zebrafish colonies and establishing techniques for their genetic analysis. He was able to produce homozygous diploid animals through artificial parthenogenesis and to generate clonal strains free of lethal mutations, which could be used for later mutational analysis. In short, the knowledge and skills from his discipline enabled Streisinger to handle a new organism within the boundaries of his discipline. However, Streisinger and his colleagues were not able to make sense of the effects of the mutations they were able to induce and map in zebrafish, since they had never dealt with multicellular organisms before. They could not grasp any phenotypic effects caused by mutation. The molecular geneticists were restricted to the generation and genetic characterization of the mutation, but they could not identify and describe any phenotypic effects. Since the embryos and larvae of zebrafish are transparent, the visual access to the phenotypic effects of a mutation facilitates grasping of the causal relations underlying development. However, while the geneticists literally saw the developmental processes within the transparent embryos, they did not "see" the effects of the induced mutation, as they did not know how normal, non-mutated zebrafish embryos develop and look. Hence, they did not and could not grasp any significant or relevant effect of the mutation during early embryogenesis. And even if the molecular geneticists learned and then knew what the normal embryo looks like from textbooks, the mutant embryo would have to be observed or maybe even dissected carefully to detect the phenotypic differences. These skills, identifying and characterizing phenotypic effects and noticing significant differences through observation or dissection, needed to be learned and trained by molecular geneticists to understand genetic functions on more complex levels of biological organization.

At this point, the knowledge and skills from embryologists and neurophysiologists entered the scene. They were able to behaviorally and physiologically describe the effects of a mutation through, for example, the fixation of embryos, measurements of neuro-muscular activities, staining sections or the execution of behavioral response tests. Without the descriptive knowledge and usage of the descriptive devices like cellular fate maps, neural wiring diagrams or staging series, the phenotypic effects of mutations could not be recognized. In other words, while the molecular geneticists were able to induce and map genetic mutations but were not able to relate this knowledge to any phenotypic effects, embryologists and neurophysiologists could identify phenotypic effects, but had no idea how genetic regulation is related to these effects or what is happening at the genetic level at all. Because of the lack of specific knowledge and skills on both sides, all the scientists involved in this new research endeavor, molecular geneticists as well as embryologists and neurophysiologists, could not grasp the relations involved in the phenomenon that they all wanted to understand. Only through the integration and combined use of the knowledge and skills from both biological disciplines were the scientists able to grasp a relation between a specific mutation and its phenotypic effect in the development of an embryo. By using mutational analysis, the scientists could determine which genetic mutation is present in a mutant, and with the skillful use of descriptive devices from embryology and neurophysiology they could literally see the phenotypic effects, and hence could grasp a relation between a mutation and a phenotypic effect. The specificities of the grasped relation, its components and structures, can be investigated by using further knowledge and skills and articulating the insights gained in the form of a new explanation. After grasping a relation between a mutation and a phenotypic effect, this relation can only be articulated if the knowledge necessary to make sense of the grasped relation is available. In the case of the neuronal necrosis mutant, the biologists could explain that the *ned-1* gene, at which the mutation was targeted, is essential to some, but not all cells of the central nervous system, because some neurons develop normally despite the mutation.[50] In order to articulate this explanation, scientists had to have the skills and tools to conduct the experiments and the necessary knowledge from genetics and neurophysiology to identify significant effects, grasp the relation between the *ned-1* gene and different groups of cells, and combine the relevant pieces of knowledge in such a way that the experimental results make sense. That is, Streisinger, Kimmel and their colleagues scientifically (and partially) understood the function of the *ned-1* gene in the early embryonic development of zebrafish.

The successful integration of knowledge and skills from both disciplines is illustrated not only in the study on neuronal necrosis mutants that Streisinger together with Kimmel and others published in 1988, but also in the adoption of mutational analysis, a technique from molecular genetics, by embryologists and physiologists. From the late 1980s on, mutational analysis became a mainstream tool in developmental biology and mutation-based studies were much more popular than classical physiological studies. This trend shows that molecular geneticists and developmental biologists actually acquired knowledge and skills from each other. In the study on the neural degeneration mutant, the developmental biologists like Kimmel acquired knowledge and skills from molecular geneticists like Streisinger, and the other way around. Their research was not a two-step study where first the molecular geneticists did their thing, and in the second step the developmental biologists did theirs,

50 See Grunwald et al. (1988).

respectively. Instead, it was an instance of a real integration of two research disciplines, together with the knowledge and skills, that was necessary in order to gain understanding of the phenomenon that everyone was interested in, the genetic regulation of embryonic development of complex organisms. This was the situation at the end the first phase of the development of zebrafish as a model organism.

What did the first phase reveal about the characteristics of scientific understanding? It showed that specific knowledge, skills, and tools were required to understand the phenomenon in question. Without the relevant knowledge from genetics and neurophysiology, the skills to conduct experiments in such a way that they allow access to the phenomenon of interest, and the tools needed for this, the respective phenomenon could not be understood. As long as the biologists lacked the required knowledge, skills and tools to conduct the research in the appropriate way, they could neither grasp any relations involved nor articulate any explanation about aspects of the phenomenon. Since the skills that scientists need to understand phenomena in a scientific way are skills to conduct scientific research, I label these skills *research skills* from now on.[51] The acquisition of research skills is closely linked to the availability of specific tools to conduct research. I am using the term "tool" in a loose sense: any material or theoretical object that can facilitate research is a tool. Examples of tools include mathematical equations, software, dissection microscopes, or cellular fate maps. If you cannot use the fate map for *C. elegans*, you also will not be able to use a fate map for zebrafish. It is the acquisition of certain research skills that enables the use of certain objects as tools in the context of research. Once you acquire the skills to read and use a fate map for *C. elegans*, you will (probably) be able to use a fate map for zebrafish.

So, in order to understand phenomena scientifically, that is, through the scientific method, scientists first need to acquire the relevant knowledge, necessary research skills, and required tools to conduct research in a way that is appropriate to understand the phenomenon in question. It is important to explicitly take these resources into account in any analysis of understanding gained in specific episodes, as the presence or absence of any of these resources might explain why phenomena were understood in some cases, but not in others. However, this is not enough to achieve the aspired comprehensive understanding of the genetic regulation of vertebrate development. The second phase of the development of zebrafish as model organism was crucial as well.

---

51 The importance of specific skills for the acquisition of understanding in science has already been recognized by, for example, Sabina Leonelli and Henk de Regt. See Leonelli, S. (2009), "Understanding in Biology: The Impure Nature of Biological Knowledge." In de Regt, H. W., Leonelli, S. & Eigner, K. (eds.), *Understanding: Philosophical Perspectives*, pp. 189–209, Pittsburgh, University of Pittsburgh Press; and de Regt, H. W. (2017). While this insight is therefore not novel, the episode around zebrafish provides additional support for it.

### 5.2.2 Generating the required material equipment and developing a research infrastructure

In addition to establishing the necessary knowledge, descriptive and manipulative tools, and the research skills to use them, the infrastructure to generate and maintain these resources, data and results of studies was set up during the second phase. This happened for two reasons.

First, following the successful integration of both disciplines in the first phase, biologists joining this research program should be equipped with the necessary knowledge and research skills to conduct successful research in the new field. Without access to the newly established combination of knowledge and research skills, new researchers in the field would not have had the chance to gain understanding of the genetic regulation of vertebrate development. Any biologist who wanted to join the new research program around zebrafish needed to have the knowledge and the research skills from molecular genetics as well as embryology and neurophysiology. Otherwise, she would have the same problem that Streisinger in the first phase was facing, before Kimmel and Westerfield joined the research project. Moreover, in order to learn the required knowledge and train the necessary research skills, scientists needed supervisors or peers who could teach them, who already possessed the knowledge and research skills from both biological disciplines. It would not have helped scientists who wanted to join the new research program if they had been trained by "pure" molecular biologists or embryologists, as they could not teach all the knowledge and research skills required for this specific research. That is, in order to do successful research on and understand the genetic regulation of vertebrates, scientists needed to build up a new research community, later called developmental genetics, in which the required expertise is maintained and can be shared with new colleagues.

Yet there was a second and not less important reason for establishing a new research infrastructure. While the ultimate goal was to understand the genetic regulation of vertebrate development in general, individual studies focused only on specific genes, their interaction and the resulting phenotypes, like the study from Streisinger and Kimmel on the function of the *ned-1* gene for the development of the nervous system. This limitation of individual studies is due to the complexity of the phenomenon being studied. Different mutations affect various structures or processes at different stages during the development, and these various genetic activities cannot be studied in only a few experimental studies. In other words, if biologists really wanted to understand the genetic regulation of vertebrate development in general, they would have to study the effect of every gene at any stage during embryonic development on any structure of the embryo. Not only would this research require a lot of time and resources, but it could not be conducted at all after the first phase of the introduction of zebrafish. The reason was that no one knew which genes are involved in developmental processes. Before one could study which effects a gene has on developmental processes, it would have been helpful to know which genes are involved in developmental processes at all. Some genes might not be involved in developmental processes; given the complexity of the envisioned research, some narrowing would have been helpful.

*The Big Screen* provided this guidance. By randomly inducing some mutations and only looking at the phenotypic effects, genes that participate in developmental processes could be identified. That is, biologists first randomly generated many different mutants and grouped them together according to similar or almost identical phenotypic traits. Only after the grouping did the biologists analyze the mutation that took place in the respective mutants. Through this method, it was possible to identify many mutations that somehow affect zebrafish embryogenesis. Now that researchers knew which genes are involved in developmental processes and had the mutant strains with identified mutations, they could start to study the actual function of the respective genes. Thus, the necessary infrastructure of the zebrafish community was established. Several new labs were founded that focused on specific mutant strains and associated developmental processes. Individual researchers conducting these specific experiments achieved understanding of the mechanisms investigated and shared their data and results with the whole community. Thus, colleagues could access the information from the individual studies and comprehend the results gained about the investigated mechanisms even though they did not conduct the experimental study themselves, and they could use the results from other studies in their own research if that seemed appropriate.[52] That is, without the generation of the various mutant strains in *The Big Screen*, the biologists working with zebrafish would have lacked the necessary material, the mutant strains, to analyze the genetic mechanisms underlying vertebrate development.

The second phase in the establishment of zebrafish as a model organism provides two further important aspects for scientific understanding. In order to understand specific phenomena, scientists need a functioning research infrastructure to share information and new insights as well as the necessary material equipment to conduct specific studies. Scientists use the knowledge that their colleagues generate and the methods they implement by applying them to understand the phenomenon

52 I explain in more detail in chapter six how it is possible to achieve understanding by receiving an explanation by testimony and not by conducting a certain experiment. Scientists are able to understand the phenomena their colleagues have researched by reading their publications or talking to them in person if they grasp the relations of the phenomenon presented in an explanation, construct relations between the information contained in the explanation and their knowledge, and draw further inferences, which have not been available to them before they received the explanation. Again, this is different from merely knowing an explanation, which does not enable scientists to put the results from other studies to use in their own research.

they themselves are researching. Additionally, results and insights gained in other studies can enable biologists to grasp new relations they might not have been aware of without the additional information from other experiments. Communication between scientists of a discipline is crucial for gaining scientific understanding and for making scientific progress. Therefore, it was essential to establish the necessary infrastructure to ensure communication among scientists and the availability of material necessary to conduct studies. You cannot understand a phenomenon if you have the necessary knowledge and relevant research skills and tools, but lack the material to work with, to actually conduct a study in which you apply and use the knowledge and research skills you possess. If you want to understand the function of a specific gene for the development of the nervous system, for example, but you do not have a mutant strain that lacks precisely this gene, you will not understand the function of this gene for the development of the nervous system.

So, the first phase showed that scientists need specific knowledge, research skills, and tools to understand a certain phenomenon scientifically. This finding is important, as it indicates that any analysis of individual cases of understanding has to consider the knowledge, research skills and equipment that was present or required for that specific case. Phase two highlighted that, additionally, scientists need an appropriate research infrastructure that ensures the distribution and maintenance of information and insights gained in individual studies as well as the required material equipment. The third phase, to which I now turn, demonstrates the importance identified in phase two of functioning communication among researchers and, furthermore, points to an additional feature of the manifestation of scientific understanding, its iterative nature.

### 5.2.3 The iterative manifestation of scientific understanding

In the context of the third and final stage of the development of zebrafish as a model organism, the study on the *oep* mutant is a further example of how biologists acquired understanding of a specific function of one gene in embryonic development. Before Gritsman and colleagues initiated their studies on Oep, some knowledge about the function of Nodal and Oep in embryonic development as well as phenotypic effects caused by respective mutations had already been established through other studies.[53] However, it was not clear with which receptors and pathways Oep interacts, and what exactly the relation is between Oep and phenotypic effects. Importantly, the research on Nodal on the one hand and Oep on the other was not yet related. Studies on the effects of Nodal and Oep in zebrafish, which revealed all the insights I just mentioned, had been conducted independently from each other.

53 See Gritsmann et al., p. 121.

The studies by Gritsman and colleagues changed that. From the very beginning, they were interested in the role of *oep*. Since the *oep* gene is expressed maternally as well as zygotically, the researchers generated embryos that lack both maternal and zygotic Oep. These mutants were called MZ*oep* mutants. When these mutants were generated, the biologists recognized that "MZ*oep* mutant embryos are very similar to double mutants for *sqt* and *cyc*, two zebrafish *nodal* related genes."[54] Only due the observed similarity between the two mutants was it possible for the biologists to grasp the relation between Oep and the Nodal signaling pathway. Through this observation, the biologists had epistemic access to the relation between Oep and Nodal signaling. Only now did they have reasons to assume that there is a connection. Importantly, grasping a relation between the two genes is something more than and distinct from merely noticing the similarity of two mutant phenotypes. Seeing the similarity was necessary but not sufficient for grasping the relation between the genes. It might have happened that the biologists saw and recognized the similarity of the phenotypes, but were unable to grasp the relation between Oep and Nodal, for example if specific knowledge about characteristics of the two genes had been missing at that time. Evidently, Gritsman and her colleagues were able to grasp a relation between Oep and Nodal only because they possessed the necessary knowledge about Oep and Nodal that had been established in other studies, and the required research skills and tools to become aware of the relation. Without the necessary resources to generate the MZ*oep* mutants and then recognize the similarity between the different mutants, it would not have been possible to grasp the relation. That the biologists had the required knowledge, research skills and tools was only due to the development of zebrafish research community over the previous decades, grounded in the integration of molecular genetics and developmental biology.

Gritsman and colleagues grasped the relation between Oep and Nodal, but they did not yet understand it. To arrive at an understanding of the function of *oep* in embryonic development, they still needed to articulate the relation in the form of an explanation of the role of Oep in Nodal signaling. This was not possible on the basis of the available knowledge and the observed similarity of the two mutants. The biologists knew that something was going on between the proteins Oep and Nodal, but they did not know how the proteins interact. Therefore, the biologists could not yet explain why and how Oep and Nodal interact or why the MZ*oep* mutants and the double mutants for *sqt* and *cyc* look so similar, because they had no epistemic access to the details of the relation. To articulate the grasped relation in an explanation, further cognitive work was necessary. Based on the already available knowledge, the biologists reflected on the observed similarity of the different mutants and considered possible reasons for it by performing abductive reasoning. They were looking for the most likely explanation of the similarity and had the idea that Oep and Sqt/

54 Ibid. p. 122.

Cyc (Nodal) might act in a common pathway. If this is the case, it would explain why the same phenotypic effects can be observed when one of the two components is missing. In both cases, the pathway would not function properly and would lead to identical effects. In general, molecular signaling pathways refer to processes by which a chemical or physical signal is transmitted through a cell as a series of molecular events, which ultimately result in a cellular response.

Assuming that Oep and Sqt/Cyc (Nodal) do act in a common pathway, how and where in the pathway do they act? Here, knowledge generated by other studies came into play. Taken into account that "Oep acts cell autonomously [...] whereas Nodal signals can act nonautonomously [...] suggested that Oep is required for cells to receive Nodal signals."[55] Still, this was only a hypothesis for which the scientists wanted supporting evidence. Through further counterfactual reasoning, they had the idea that, if Oep is not required for Nodal signaling, MZ*oep* mutants would be rescued by injecting mRNAs encoding Sqt, Cyc, or mouse Nodal as replacement for Oep, and they devised the respective experiment. In other words, if the idea that Nodal signaling necessarily requires Oep is false, the injection of Nodal in the absence of Oep would lead to a normal development of the mutants. Since the mutants were not rescued through this procedure, the biologists obtained the evidence that Oep is indeed essential for Nodal signaling and that Nodal signaling does not take place without Oep, and they could explain the role of Oep during embryogenesis with this observation. The use of further available knowledge from other studies and research skills enabled the biologists to grasp and explain this crucial detail of the relation between Oep and Nodal, the necessity of Oep.

However, the biologists did not yet achieve the understanding of the function of Oep in the Nodal signaling pathway to which they aspired. It was still not understood exactly which step in the Nodal signaling pathway requires Oep. Again, results obtained by other research groups were crucial to understand this aspect. Since other studies suggested but did not definitively show that Nodal signaling is mediated by a pathway that might involve the ActRIB receptor and the Smad2 transcription factor, Gritsman and colleagues wanted to test whether "Oep is essential for the responses to these factors."[56] At this step, the counterfactual reasoning process was that if Nodal signaling is transmitted by the indicated pathway, and Oep is essential for Nodal signaling, the pathway will not be activated if Oep is absent. In other words, Oep activates this particular pathway and the ActRIB receptor as well as the Smad2 transcription factor act downstream of Oep in the causal pathway. If this is the case, it should be possible to activate the Nodal signaling pathway at a subsequent step, one that follows the activation by Oep, in the MZ*oep* mutants by

55 Ibid. p. 125.
56 Ibid. p. 125.

injecting already activated ActRIB and Smad2 in the mutants. The impossible activation of the Nodal signaling pathway by Oep in the mutants would be replaced by an activation of the downstream components outside of the mutants, which are then injected. Since the mutants were rescued as a result of this injection, the hypothesis could be confirmed and the biologists arrived at a new understanding of Oep "as an essential component of Nodal signaling [...] that allows Nodal to activate its downstream signaling pathway."[57] They had grasped and articulated where in the pathway Oep executes its function, how Oep and Nodal are related.

This example, the process of understanding the function of Oep for Nodal signaling in vertebrate development, demonstrates the complexity of the manifestation of scientific understanding. Gritsman and colleagues had to have access to the required knowledge already generated by other studies, possess the relevant research skills and tools to conduct research that enabled them to grasp the relation between Oep and Nodal in the first place. Only subsequently could they detect the components and structure of the relation more precisely, like the exact role of Oep as an essential cofactor for Nodal signals and its position in the causal relation, as allowing for the activation of the downstream signaling pathway. Without the pieces of knowledge about *nodal* and *oep* gained in other studies, Gritsman and colleagues would not have been able to understand their relation. Even if they had recognized the similarity between the *oep* mutant and the other mutant lacking *nodal* and concluded that the genes must somehow be related, they would have had no chance to understand how the genes are related without some preexisting knowledge about these genes. If you do not know that Oep acts cell autonomously, while Nodal can act nonautonomously, you will never have the idea that Oep might be necessary for cells to receive Nodal signal, not to mention that this might actually be the case. Without some minimal or hypothetical knowledge about aspects of the phenomenon to be understood, scientists would not have any starting point for their research, would have no idea where to start or which hypothesis could be tested first. If scientists have some knowledge to start with, they then need the research skills, tools, and material to do the research that they hope will allow them to understand the phenomenon. In the case of the research on the *oep*-mutant, Gritsman and colleagues had knowledge about Oep and Nodal from other studies, had the research skills and tools to induce and map mutations in the *oep*-mutants as well as to identify significant phenotypic effects and, last but by far not least, they had the material, that is, the specific zebrafish mutant strain to work on.

All of these resources, the knowledge, research skills, tools, and material were provided from the established and functioning research infrastructure around zebrafish. If Gritsman and colleagues had not been trained in the research skills and

57 Ibid. p. 129.

the use of tools from the newly established field of developmental genetics, comprising the formerly separate fields of molecular genetics, embryology and neurophysiology, had not have access to the results gained in other studies on Oep and Nodal, and had not have the *oep*-mutant, which was identified in *The Big Screen*, they would not have been able to do the research they actually did and to gain the understanding of the function of *oep* for vertebrate development that they did acquire. The team around Gritsman needed the broader research infrastructure that provided them with all the resources they needed in order to acquire scientific understanding. It was *The Big Screen* that made the identification of the various mutations, the founding of new labs focusing on different mutants and hence parallel research on different mutations and their effects on embryonic development, which resulted in the discovery of novel insights published and made available for other scientists who might need these insights for their own research, possible. In order to get understanding of a specific phenomenon, scientists must be part of an appropriate infrastructure. In the example of the function of *oep* in vertebrate development, Gritsman and colleagues were part of an infrastructure appropriate for understanding this phenomenon.

So, the example of the research conducted by Gritsman and her team within the third phase of the establishment of zebrafish as a model organism corroborates my claims that the first and second phase of the episode and the resources established during these phases are necessary for the scientific understanding of the genetic regulation of vertebrate development. However, the studies on *oep* and *nodal* reveal an additional crucial aspect of the manifestation of understanding. I argue in chapter 4.3 that the ability to understand a phenomenon manifests in the process of grasping relations the phenomenon stands in and articulating these relations in the form of explanations. As the research on the *oep*-mutant shows, biologists did not grasp all details of the relation in question at once and then articulate an explanation of this relation. That is, the biologists did not first grasp everything there was to grasp and then articulate this in an explanation, as the characterization in chapter 4.3 might suggest. Rather, grasping relations and articulating explanations – manifesting understanding – is an iterative process. Grasping and explaining depend on each other. Let me elaborate this idea again with the research on *oep*.

When the scientists generated the *oep*-mutants, the process of understanding began with grasping the similarity relation between the phenotypes of the generated *oep*-mutants and the double mutant for *cyc* and *sqt*. Based on the knowledge of which genes the two mutant strains are lacking and the observation that both mutant strains have a similar phenotype, the biologists reasoned that the genes missing in one of each mutant type must be related. Only due the observed similarity of the different mutants was it possible for the biologists to grasp a relation of the respective genes. The biologists had reasons to assume that there is a relation, but they did not yet understand this relation. To gain understanding of what is going on and how the genes interact, the biologists were looking for reasons for the observed similarity of the different mutants. They were looking for an explanation of the similarity and had the idea that the proteins encoded by the genes might act in a common molecular pathway. If this is the case, it would explain why the same phenotypic effects can be observed when one of the two components of the same pathway is missing. In both cases, the pathway would not function properly and would lead to identical effects. So, the biologists articulated a first hypothetical explanation: the zebrafish mutants that lack *oep* or *nodal* have a very similar phenotype because the proteins encoded by the two genes act in a common molecular pathway.

They understood that the two genes are related, but they did not know whether the genes really act on a common molecular pathway and if so, how exactly they interact. To answer these questions, the biologists referred to the results and knowledge about features of these genes gained in other studies. At this stage, the biologists grasped that the insights from these other studies are related to the function of *oep* for embryonic development. The integration of this additional knowledge "suggested that Oep is required for cells to receive Nodal signals."[58] This is already a more concrete conception of the relation of the genes, more concrete than just saying the genes somehow act on some common pathway. The biologists arrived at the following hypothetical explanation: in normal, non-mutated fish, *oep* has an important function in embryonic development, because it activates the Nodal signaling pathway by which several early embryonic developmental processes are regulated.

However, this was still only a hypothetical explanation. The scientists wanted supporting evidence to ensure that they understood the function of *oep* correctly, that this explanation represents the relation of Oep and Nodal correctly. In a third step, the biologists designed and conducted several experiments to determine whether Oep is indeed necessary for Nodal signaling. These experiments did show that the scientists were right, that Oep is indeed essential for Nodal signaling. The biologists could confirm their hypothetical explanation. Nonetheless, before conducting the additional experiments in the third step, the biologists could not know whether their articulated hypothetical explanation was correct. That is, they could not know whether they already understood the function of *oep*, or rather misunderstood it. It could have happened that the experiments in which the biologists tested the hypothetical explanation falsified this explanation, instead of confirming it. You do not know this in advance, which is why you test your explanations. If the hypothetical explanation would have been falsified, the biologists would have realized that they had misunderstood the function of *oep*, that is, that their explanation did not represent relations of the phenomenon. When phenomenon and explanation conflict, this motivates scientists to work out and articulate an alternative explanation, to understand the phenomenon in a different way in which the conflict

58 Ibid. p. 125.

dissolves. This observation nicely fits Michael Polanyi's idea, presented in section 4.2, that understanding brings our language and the world in line, establishing coherence among language and perception.

Ultimately, the iterative process of grasping and explaining, of taking various pieces of knowledge into account by performing, in this case, abductive and counterfactual reasoning and testing generated hypothetical explanations by employing research skills for the intervention or manipulation in the model organism finally enabled the researchers to articulate the following explanation, to arrive at the following understanding, of the function of Oep in vertebrate development: Oep is required for vertebrate embryogenesis, because it activates the Nodal signaling pathway by which germ layer formation, organizer development, and the positioning of the anterior-posterior axis are regulated.

## 5.3 Understanding the genetic regulation of vertebrate development scientifically

How did the research on zebrafish allow for scientific understanding? Analyzing this episode from scientific practice revealed three important and related insights.

First, biologists could understand the genetic regulation of embryonic development of complex organisms because they had the necessary knowledge from molecular genetics as well as from developmental biology, viz. embryology and neurophysiology, as well as the research skills and tools from both disciplines. These resources enabled biologists to grasp relations between genetic activities and developed phenotypes in a given experiment. Material skills, like the skills to map and isolate genes or to fix and dissect embryos, to name just a few, were necessary for understanding, because their application enabled the scientists to carve out, to isolate, relations that are part of the phenomenon of interest. Without the possession and use of these material skills, scientists would not have been able to investigate any phenomenon they did not yet understand. They would never have gained epistemic access to the phenomenon of interest. When a relation was grasped, when scientists became aware of it after it had been isolated with the aid of research skills, they tried to make sense of that relation by figuring out its details. In the case presented here, abductive and counterfactual reasoning was used to hypothesize what this relation might look like, what its details are, and to articulate a tentative explanation on the basis of this reasoning process. The hypothesized explanation was then tested in subsequent studies, for which additional research skills as well as additional knowledge may have been required. Intervening in the grasped relation again allowed for epistemic access to more details of the relation. When the scientists conducting a specific study grasped no more details of a relation and articulated and tested all the aspects that were grasped, they formulated a final explanation – for the study in question – and gained understanding of the phenomenon of interest to the extent that was possible in the setting of the study. The history and use of zebrafish as a model organism to understand the genetic regulation of vertebrate development shows that scientific understanding of empirical phenomena requires the availability and use of relevant knowledge, research skills and related tools.

Second, the episode demonstrates that achieving scientific understanding is an extremely complex and demanding process that requires and is influenced by an appropriate context or environment. All the participating scientists in this project had the goal of understanding the genetic regulation of vertebrate development, but realizing this goal required a huge community. Besides establishing and learning the necessary knowledge and research skills, an infrastructure to secure the communication and distribution of theoretical as well as material resources needed to be implemented. These resources cover zebrafish mutant strains, material tools to work on the mutants, as well as knowledge that was gained in individual studies. Since the genetic regulation of vertebrate development is a very complex phenomenon, single scientists or groups of scientists will never be able to understand this phenomenon without the results and support from other research groups. Because the research community working on zebrafish split up into several research groups working on specific genes or specific developmental processes, it was necessary to formulate what was understood about an aspect of the genetic regulation in the form of an explanation. The acquired knowledge or explanation could then be communicated to others, who can scrutinize or also use the shared knowledge for their own research. Each study contributes insights about parts of the phenomenon, but only as a community with a functioning infrastructure can the genetic regulation of vertebrate development be scientifically understood in its entirety.

The third important insight provided by this episode is the stepwise and iterative manifestation of scientific understanding. The scientists did not do their research, then grasp every relation or all the details of a relation of a phenomenon at once, and then articulate the one "final" explanation for the respective study. Instead, manifesting scientific understanding of a phenomenon is an iterative process of applying scientific methods, grasping relations that were carved out by the method, articulating the grasped relation in an explanation through reasoning or additional research, which again enables grasping further relations or details of an already grasped relation, the successive articulation of an additional or more precise explanation, and so on. This goes on until the scientists decide that, for some specific study, they have understood the phenomenon sufficiently for the time being and publish their results. And the iterative manifestation of scientific understanding parallels the stepwise combination of interventions in the model organism that Meunier emphasizes.

Generally, my analysis of this scientific episode matches the views from Ankeny & Leonelli and also from Meunier concerning model organisms and their use in scientific practice. Recalling the quote from the beginning of this chapter, the characteristic feature of model organisms, according to Ankeny & Leonelli, is its use as a model for organisms as wholes and for gaining an integrative understanding of intact organisms with regard to their genetics, development and physiology. These are exactly the goals of the biologists working on zebrafish. They wanted to understand the genetic regulation of vertebrate development as a whole. Understanding the more particular phenomena involved, like the effect of injecting certain mRNAs in specific mutants, were necessary instrumental steps in the process of understanding the phenomenon that was ultimately of interest. In all the studies on the various zebrafish mutants identified in *The Big Screen* and numerous experiments within each study, the scientists operated the mechanism and thereby manipulated the development, as Meunier expresses in the quote I present in section 5.1.4. Through the stepwise combination of interventions, the mechanism was carved out and analyzed, and thereby understood. That the biologists could carve out the mechanism by intervention and manipulation was only possible because an appropriate research infrastructure was established. As I already said, this infrastructure secured the communication among scientists and the distribution of the necessary material equipment, research skills and tools and the knowledge acquired in the studies. Through carving out the genetic mechanism in the mutants, it could be grasped and articulated in an explanation, which can then serve as a basis for the construction of theoretical models of the mechanism in other organisms, according to Meunier. This is in line with the ultimate goal of Gritsman and colleagues in studying Oep. They did not want to understand the effects of the presence or absence of Oep in certain mutants, but rather the general function of Oep in normal, unmanipulated embryonic development.

In sum, the episode from the research around zebrafish supports my view developed in the previous two chapters. In chapter four, I argued that understanding is an ability to make sense of a phenomenon through aligning experience and the knowledge stored in the respective language an individual uses. Scientists working on zebrafish did exactly this, they made sense of the causes of embryonic development by bringing their "experience" (in the case of science, observations of embryos or genetic data) in line with what they already knew about genes and embryonic development. Since the scientists involved articulated and published explanations about aspects of the genetic regulation of vertebrate development that they understood, the episode also sustains the claim I defend in chapter three, namely that scientific understanding requires explanation. Beyond that, the analysis of this episode shows that, first, scientific understanding of some phenomenon requires relevant knowledge, specific research skills and tools. Second that an appropriate community ensuring the generation and distribution of needed (material) equipment is necessary. And finally, the episode highlights that the manifestation of understanding is an iterative process, consisting of several subsequent steps of grasping some relation or aspects thereof and articulating what has been grasped in a (tentative) explanation. These are the three important findings I take into the next chapter.

So, I maintain that the episode analyzed here provides crucial insights about the nature of scientific understanding and the way scientists achieve scientific understanding of phenomena. Am I thereby making the same problematic move I accuse other philosophers of science of at the beginning of this chapter, generating claims about scientific understanding by looking at one particular discipline? Is it not the case that the episode about zebrafish can only reveal insights about understanding gained in biology, about "biological understanding", or even more narrowly, about understanding gained in developmental genetics, the discipline that developed together with the establishment of zebrafish as a model organism?

I hope not, as I follow Hasok Chang in

> "seeing the history-philosophy relation as one between the *concrete* and the *abstract*, instead of one between the particular and the general. Abstract ideas are needed for the understanding of *any* concrete episode, so we could not avoid them even if we only ever had one episode to deal with. [...] Any concrete account requires abstract notions in the characterization of the relevant events, characters, circumstances and decisions. If we extract abstract insights from the account of a specific concrete episode that we have produced ourselves, that is not so much a process of *generalization*, as an *articulation* of what we already put into it."[59]

In this chapter, I looked at one concrete episode. In the next chapter, I abstract away from this concrete episode and develop my account of scientific understanding by taking the three insights gained from the concrete episode into account.

59 Chang (2012), p. 110. I used the exact same quote already in the first footnote of this chapter where I clarified my use of the term "episode".

# 6. Grasping and explaining – an account of scientific understanding

What is scientific understanding and how is it achieved? In the previous chapters, I argue that understanding requires explanation and that understanding should be conceived as an ability, rather than a type of propositional knowledge, which manifests in the process of grasping relations and articulating these in the form of explanations. Through analyzing the episode of how biologists gained understanding of the genetic regulation of vertebrate development, I show that scientific understanding necessitates propositional knowledge as well as further research skills and appropriate equipment for its manifestation. Now, I bring all these lines of thought together to provide a novel account of scientific understanding. In section 6.1, I present and explain the scope and conditions of the 'Grasping and Explaining-Account of Scientific Understanding', the GE-account in short. In section 6.2, I elaborate the advantages of the GE-account in contrast to the accounts of understanding developed by Henk de Regt, Kareem Khalifa and Finnur Dellsén, which are introduced in chapter two.

Importantly, all the claims I am making here are supposed to hold for understanding of phenomena achieved in science through scientific methods and practices. I stay agnostic about the extent to which the GE-account of scientific understanding developed here may also apply to other types of understanding in other contexts.[1]

## 6.1 The GE-account of scientific understanding

I argue that scientific understanding is an extensive, complex cognitive ability that individual scientists possess and that is manifested in the process of grasping relations between pieces or bodies of knowledge and investigated phenomena, between

1 Therefore, I will occasionally talk about understanding without the qualifier *scientific*, but this then also refers to scientific understanding. In cases where I discuss other forms of understanding or understanding in general, I will clarify this explicitly.

several phenomena or within parts or aspects of a phenomenon, among its parts, that scientists were not aware of before, and articulating these relations in form of explanations. Explanation is a central and necessary element of understanding, as it makes the recognized relations between knowledge and phenomena or within phenomena comprehensible and revisable for the scientist. In order to manifest this ability, to actually understand some aspect of a phenomenon, the availability and usage of propositional knowledge, research skills, as well as an appropriate equipment are required. If one of these elements is missing, it will not be possible for a scientist to achieve understanding.

The GE-account of scientific understanding that I elaborate and argue for in this chapter entails necessary and sufficient conditions for acquiring scientific understanding and takes the following form:

> A scientist *S* has scientific understanding of an empirical phenomenon *P* in a context *C* if and only if
>
> i. *S* grasps (details of) relations that *P* stands in and articulates these relations in the form of new explanations of (aspects of) *P (manifestation condition)*,
> ii. *S* possesses and uses (material) equipment, relevant knowledge and research skills provided by *C* and required for understanding *P (resource condition)*, and
> iii. *S* is a member of a scientific community that enables *S* to understand *P* and parts of that community approve *S*'s understanding of *P (justification condition)*.

This means that understanding is the ability to make sense of a phenomenon through using knowledge, equipment and research skills that are at a scientist's disposal in a reasoning and research process. A result of this reasoning process is a new explanation. The possession of required resources, covering (background) knowledge, equipment, and specific research skills, is a necessary precondition for scientific understanding. These resources, which are acquired, learned and trained by scientists during their education and practice, allow for the grasping of relations of a phenomenon, which have not been known by the subject before she started to reason about the phenomenon in question. And the relations that have been grasped are then articulated in the form of an explanation, in order to identify and specify the nature and aspects of the relations. Additionally, explanations allow for an assessment of the acquired understanding of an individual scientists by other members of the respective research community. When a scientist has performed this process, when her scientific understanding has become manifest, she will have produced new knowledge. Namely, the knowledge of an explanation that was not known to her before, and possibly not by anyone else. Scientific knowledge is a product of scientific understanding, but it is not identical to understanding.

Understanding is the ability to generate new knowledge, which exceeds the mere possession of knowledge. In what follows, I elaborate the GE-account of scientific understanding, its scope and conditions, in detail.

### 6.1.1 Clarifying the scope of the GE-account

Let me start with the scope of the GE-account, before I turn to the conditions. First of all, understanding requires a subject and an object: Someone understands something. The subject of understanding I am focussing on is the individual scientist. I will not analyse what some kind of collective understanding amounts to. Although there is already some work on understanding on the level of groups,[2] I focus on the smallest unit of understanding, which is the individual scientist. So, the individual scientist is the subject of understanding, the one who understands, and an empirical phenomenon is the object of understanding, the thing that is understood. Since I want the GE-account of scientific understanding to accommodate as many scientific disciplines as possible, I adopt a very broad notion of what a phenomenon is and follow Hans-Jörg Rheinberger, who takes phenomena to be epistemic things that "embody what one does not yet know."[3] Objects, structures, events, processes, or mechanisms belong to the phenomena that scientists investigate. Hence, the GE-account of scientific understanding does not cover understanding of theories, explanations, models, or other representations used in science. While the GE-account requires the articulation of an explanation for the manifestation of understanding, for which theories may be necessary, neither explanations nor theories are the objects that scientists want to understand. Ultimately, scientists want to understand phenomena in the world, ranging from quantum phenomena, chemical reactions, genetic interactions, to geological or social phenomena, to mention just some examples. Theories, explanations, models and the like are necessary means to reach this goal. Whether the understanding of these representations can be captured by the GE-account as well, or whether and in which ways the understanding of representations differs from understanding empirical phenomena, are questions that I have to leave unanswered.

While the GE-account focusses on the understanding that individual scientists achieve, it does pay attention to the crucial role that the context plays for the understanding that any scientists can possibly achieve. Every scientist is situated in a specific disciplinary, historical, technological, and social context that has an influence on the understanding she may be able to achieve. I illustrate the context-

---

2 For an analysis of group or collective understanding, see for example Boyd, K. (2019). "Group understanding." *Synthese*, 198 (7), pp. 6837–6858, DOI: 10.1007/s11229-019-02492-3.

3 Rheinberger, H.-J. (1997), *Toward a history of epistemic things. Synthesizing proteins in the test tube*, Stanford (CA), Stanford University Press, p. 28.

sensitivity of scientific understanding with the episode of the research on zebrafish in chapter five and argue for the context-sensitive nature of understanding in chapter 4.2. In chapter four, I rely on the work from Michael Polanyi on the relation between articulate intelligence (formal or explicit reasoning) and inarticulate intelligence (abilities like understanding). According to Polanyi, scientists need explicit formal frameworks, covering propositional knowledge and theories, in order to understand phenomena, because the respective disciplinary formal framework systematically stores a huge amount of information potentially relevant for the phenomenon in question, without which it would be much more difficult or even impossible to understand the phenomenon. Contemporary scientists do not start from nowhere when initiating a new research project. Instead, they rely on the well-established and confirmed conceptual frameworks that their predecessors established over decades and centuries. While these conceptual frameworks are never immune to revision, they nevertheless function like glasses through which the new phenomenon is viewed in a specific way. This claim is illustrated with the episode from biology in chapter five. Before molecular biologists and developmental biologists joined forces and established the new field of developmental genetics, none of the involved scientists were able to understand the phenomenon in question, that is, the genetic regulation of vertebrate development. This was partially due to insufficient formal conceptual frameworks of both biological disciplines. Molecular biologists lacked the concepts and language to talk and think about developmental phenomena, and vice versa, the conceptual framework that developmental biologists used so far did not cover phenomena at the genetic level. Researchers participating in the study of genetic regulation of vertebrate development had to revise their respective formal frameworks, in this case through an integration of two already existing ones.

So, in general, the community to which an individual scientist belongs significantly affects the understanding of a phenomenon that she may achieve. This is why the GE-account covers cases of scientific understanding of an empirical phenomenon *P* that an individual scientist *S* gains in a context *C*. Figure 1 depicts this basic idea.

The various dimensions in which understanding is affected by the context are captured by the *resource condition* and the *justification condition* of the GE-account, which I spell out in sections 6.1.3 and 6.1.4. Before I do this, let's have a look at the first condition of the GE-account, the *manifestation condition* of scientific understanding.

*Figure 1: An individual scientist S understands an empirical phenomenon P in a context C.*

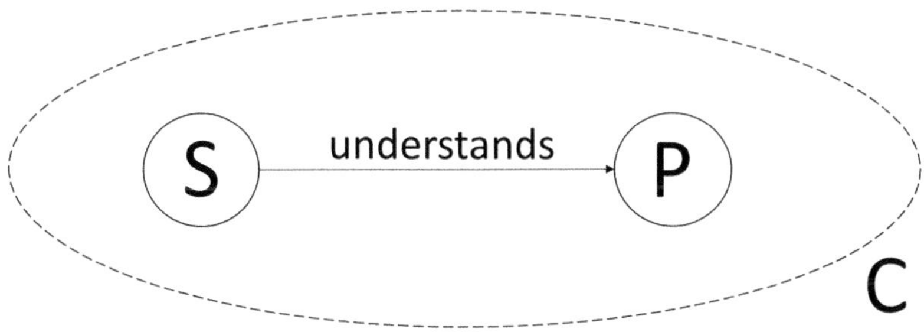


## 6.1.2 The *manifestation condition*

As I just mentioned, the GE-account only captures scientific understanding of empirical phenomena. However, in this first condition I talk about relations that *P* stands in and about aspects of *P*. I explain why it is important to talk about relations and aspects of *P* in the context of the manifestation process of understanding, the process of grasping relations and articulating them in the form of explanations.

### 6.1.2.1 Grasping relations reviewed

Let me start with relations. The concept of relations, or dependency relations, is central in the debate on understanding in general, not only on scientific understanding. Despite there being many points of contention, there is some basic consensus or some shared intuition that understanding is something like "seeing how things hang together".[4] And things hang together through some kinds of relations. Stephen Grimm, for instance, states that dependency relations are the objects of understanding and illustrates this claim with two examples. If we want to understand why a cup of coffee spilled, we must grasp the relation of the spilling of the cup to the nudging of the table that caused the spilling. If we want to understand the US House of Representatives, we have to grasp various dependency relations among its elements, its composition, its powers, and procedures.[5] Christoph Baumberger argues that dependency relations can be different in kind, ranging from relations that hold between parts or aspects of the phenomenon that is to be understood (such as causal, probabilistic, mereological, supervenience, and teleological relations) to relations among elements of a body of information through which a phenomenon is understood (such as logical, conceptual and explanatory relations).[6] And alterna-

4 See for example Baumberger, Beisbart & Brun (2017), p. 12.

5 See Grimm (2017), pp. 214ff.

6 See Baumberger (2011), p. 79.

tively, according to Kareem Khalifa's model of understanding, scientists grasp the *explanatory nexus* of a phenomenon, where the *explanatory nexus* contains correct explanations of *p* as well as the *relations* between those explanations. Note that Khalifa does not suggest an ontic view of explanation, i.e. that explanations are mind-independent things in the world. Instead, he explicitly states that he is noncommittal about the nature of explanation. Khalifa allows that the notion of explanation can be identified with the notion of explanatory information and he does occasionally talk about explanatory factors or features. Explanations can represent mechanisms, causal structures, but also non-causal, contrastive or probabilistic relations.[7]

In sum, the concept of relation is omnipresent in the debate about understanding. To my knowledge, every scholar involved refers somehow to (dependency or other kinds of) relations, relationships, connections or ties when talking about understanding. However, the agreement does not go much further. Regarding the question what kinds of relations can or must be grasped for understanding, various answers and views can be found. While the disagreement about the nature of relations that need to be grasped for understanding may affect some issues regarding different types of understanding, it does not affect the GE-account of scientific understanding. As the various scientific disciplines understand various different kinds of phenomena, these different and diverse phenomena (probably or reasonably) stand in various different and diverse relations towards various different kinds of other things. These may include relations among the parts of a phenomenon (internal relations), between the phenomenon and its parts (for example mereological or grounding relations), among different phenomena (for instance causal or statistical relations), and theories or bodies of knowledge that are taken to represent phenomena (representational relations). What kinds of relations a specific phenomenon, its parts or aspects stand in can only be analysed in the individual case of understanding that phenomenon. Note that all the kinds of relations I am mentioning here are only meant to be examples. I do not intend to provide any representative or even complete list of kinds of relations that phenomena can stand in. Fortunately, as the GE-account is supposed to cover many (ideally all) scientific disciplines, the precise nature of the grasped relations is irrelevant for the abstract account. As the basic consensus only demands that some kinds of relations must be grasped for understanding, it is not necessary for the GE-account to specify any kind of relation that needs to be grasped for understanding any of the diverse phenomena that are the objects of scientific research. Again, it must be investigated case by case which relations a specific phenomenon stands in. In the episode from the research on zebrafish, for instance, causal relations among

7 See Khalifa (2017b), pp. 6ff.

genes were of interest, and hence had to be grasped.[8] That a phenomenon may stand in various different kinds of relations is also the reason why I do not only talk about phenomena, but also about their parts and aspects. Referring only to phenomena and their parts may lead to the impression that I put an emphasis on mereological relations.[9] While mereological relations may be important for understanding a phenomenon, I want to allow all possible kinds of relations to be grasped if required. Hence, I occasionally refer to the more abstract term 'aspects' of phenomena. I return to the differentiation between phenomenon, part or aspect in section 6.1.2.3, after addressing the two processes that make up the manifestation of understanding. One of these is grasping.

The concept of grasping is almost inseparably tied to the concept of relation, at least in the debate on understanding. Again, almost every scholar in the debate agrees that to understand a phenomenon, scientists have to grasp relations of that phenomenon, that grasping demarcates understanding from knowledge, but there is no consensus what exactly it means to grasp relations. I address and discuss this question what grasping is or what it should be taken to be in section 4.3.1 and conclude that the so called "naturalistic view", according to which grasping a relation amounts to recognizing this relation and being aware of it, is the most plausible option. I identify grasping as having epistemic access to a relation of the phenomenon, that a scientist establishes some connection between her mind and the world through grasping, which is the view that, for example, Michael Strevens and Alexander Reutlinger et al. hold as well.[10] Understanding is a cognitive ability, and if we want to understand some phenomenon that lies outside of our mind, in the world, we somehow have to "connect" our mind to the phenomenon. That is what we do through grasping. When a person grasps a relation, this relation somehow catches her attention, it gets into her focus. She is somehow aware that there is something interesting or relevant about the phenomenon that she wants to understand.

It happens only in the next step, after recognizing or grasping that some relation is there, that the person applies modal, counterfactual, inductive, deductive or analogue reasoning, to make sense of the relation that was just grasped. I take grasping

8 For a basic overview over philosophical discussions concerning relations, see for example MacBride, F., "Relations", *The Stanford Encyclopedia of Philosophy* (Winter 2020 Edition), Edward N. Zalta (ed.), URL = https://plato.stanford.edu/archives/win2020/entries/relations/ (last accessed April 14th, 2022).

9 It should be noted that the term 'part' itself has no ontological restriction in mereology, c.f. Varzi, A., "Mereology", *The Stanford Encyclopedia of Philosophy* (Spring 2019 Edition), Edward N. Zalta (ed.), URL = https://plato.stanford.edu/archives/spr2019/entries/mereology/ (last accessed April 14th, 2022).

10 See Reutlinger, Hangleiter & Hartmann (2018); and Strevens (2013).

to be a process that is distinct from other reasoning abilities. In my view, it is implausible to mix up grasping and other reasoning or inferential abilities, as a subject first needs to identify something to reason about. A subject cannot make modal or other kinds of inferences about something that she is not aware of in the first place. For example, I cannot understand or even reason about global climate change if I do not grasp some (potential) causes or physical laws involved in climate change. Hence, I take grasping a relation to be a process that precedes any thinking about that relation. Grasping a relation is a process that foregoes and also parallels reasoning about that relation. The subject will immediately start to reason about the relation once she grasped it, in order to make sense of how the grasped relation is involved in the phenomenon one wants to understand. That is, once I grasped that energy conservation laws may have something to do with global climate change, I begin to reason about this relation in order to make sense of it. However, merely reasoning about the relation of energy conservation laws and global climate change is no guarantee that I will make sense of that relation. This is why I do not include the notion of reasoning in the *manifestation condition* of understanding, but rather a different process.

#### 6.1.2.2 Articulating new explanations

Namely, the articulation of the grasped relation in form of an explanation is the second step in the manifestation of understanding. The articulation is necessary for clarifying and testing what exactly has been grasped, to make the grasped relation comprehensible and revisable. I argue in section 4.3 that being able to grasp relations is necessary, but not sufficient for understanding, because a scientist cannot make sense of the grasped relations without the articulation of the relations. Through grasping, she will know that something important or interesting is going on that she has not been aware of before, but she will not be able to make sense, to discover aspects, of the grasped relation. As a result, the grasped relation will remain opaque for the scientist. By articulating an explanation, the scientist can sort out and specify the aspects of the grasped relation, which can then be presented, assessed or even corrected by herself or her colleagues. In short, one can say that through the articulation of the grasped relation in the form of an explanation, a scientist combines what she has grasped with further knowledge that she already possesses, her conceptual framework, in a consistent manner. Without articulating the grasped relation in form of an explanation, scientists will not be able to understand (aspects of) the phenomenon that stands in the respective relation, because scientists can only think about any empirical phenomena by using the specialized language and terminology they learned. Only by using and applying the respective terminology, classificatory systems, or nomenclatures can scientists sort out how parts or aspects of the phenomenon are related and what kind of relation holds. Together, grasping relations and articulating these relations in the form of explanations manifests understanding.

Recall that in chapter three, I adopt an epistemic conception of explanation and allow for a plurality of models of explanation to provide understanding. I view an explanation to be a representation of relations of (parts or aspects of) the phenomenon under investigation, which provides reasons (explanans) for characteristics of (parts or aspects of) the phenomenon (explanandum). This notion of explanation is very generic, which is necessary since it is intended to accommodate the various different kinds of phenomena, which can stand in various different relations, that scientists investigate as well as the demand for an explanatory pluralism found in science. Investigations of scientific practice revealed that various forms of scientific explanations exist (e.g., causal, mechanical, unificationist, functional, model-based, contrastive, probabilistic ...), are legitimately used, and that all of them can provide scientific understanding in certain contexts, as no timeless or universal criteria for explanation (and understanding) exist. Varying kinds of explanation are grounded in different perspectives on a phenomenon due to different formal frameworks, present different relations that were grasped, and lead to an increase in knowledge, specifically in terms of its diversity, which could become relevant in diverse contexts. Hence, the GE-account accommodates my argumentation in chapter three, in which I claim that an explanatory account of understanding, an account that conceptualizes explanation as necessary for understanding, is more appropriate to capture scientific understanding in light of scientific practice. According to the GE-account, understanding and explanation are related in the sense that explanation is a necessary product of the manifestation of scientific understanding. This claim is also supported by the episode from the biological research on zebrafish, in which scientists articulated and communicated explanations of the respective (aspects of the) phenomena they were investigating in the various studies.

In this *manifestation condition* of scientific understanding, I require that new explanations are articulated. What do I mean with 'new' explanations, and is this qualification important? Being 'new' in my usage of the term means that the explanation represents a relation that has not been known to the subject that articulated this particular explanation, and hence acquired understanding, before. The explanation must be new *only* to the individual subject, i.e. my use of the term 'new' is very local and relative to the subject. Other scientists may already possess this explanation, but this would not change the fact that a specific individual, or maybe even several individuals in a research group, articulate the explanation, and therefore have gained understanding, without having heard of or read the explanation before. The possibility that other scientists may already possess a specific explanation does not downgrade the achievement of an individual who came up with the same explanation by herself without having known the explanation before. Consider the hypothetical case that two scientists, Amy and Bob, within a very large research community investigate the exact same phenomenon, asking the same research questions and, by coincidence, come to grasp the same relations and articulate identical explanations. However, due to the many members of that community, Amy and Bob did not know each other when they gained understanding, respectively. The fact that someone else on the planet might have or actually has understood a phenomenon in the exact same way does not devalue the understanding achieved by any other subject. Such constellations always appear, for example, in supervisor and student relations in academia. Students are expected to use the knowledge they gain in lectures and the skills they train during their education to solve problems (i.e. construct explanations) of phenomena they did not understand before. That their supervisors already have this understanding of the phenomena does not change anything about the understanding the students acquire by creating solutions or explanations which are new to them. This remains a great cognitive achievement.

An explanation can be new in three different respects:

1) the explanans for a known explanandum can be new,
2) a known explanans is related to a new explanandum, or
3) both the explanans and the explanandum, i.e. the whole explanation, can be new.

In the first case, a known phenomenon is explained differently, e.g. due to additional research. Khalifa provides the example of research on peptic ulcers. First, scientists thought that acid causes peptic ulcers, but it has been discovered that bacteria are the actual cause of peptic ulcers. In this case, the explanans "caused by acidity" has been replaced by "caused by bacteria", while the explanandum "the occurrence of peptic ulcers" stayed the same.[11] Therefore, the understanding of an already known phenomenon changed. In the second case, scientific understanding is acquired by applying an already known explanans to a new explanandum, i.e. a new phenomenon. For example, if one can already explain the motion of the Earth through Kepler's law, one can also explain the motion of other planets in different solar systems which were just discovered by a brand-new high-resolution telescope (new phenomena) with Kepler's law. In these cases, a different and new phenomenon is understood. Both explanans and explanandum are labelled new if a new phenomenon is discovered (e.g. the appearance of a new butterfly species) for which there has not been an explanation before and through conducting research an explanans is generated that did not already apply to any other phenomenon (e.g. this specific butterfly species evolved in this way because of the very specific environmental changes, which affect only this species due to the niche it occupies). In this case, too, a new phenomenon is understood.

11 See Khalifa (2017a) for more information about the research on peptic ulcers.

Still, why is it important to articulate *a new* explanation, and why not simply *an* explanation? This difference is important because the GE-account of scientific understanding is intended to cover understanding that is gained through conducting scientific research or scientific practice. The account is not intended to capture cases in which scientists did not perform research on a specific phenomenon by themselves, but gained understanding through receiving an explanation of a phenomenon by listening to the testimony of colleagues or reading about a new explanation in a journal article published by a different research group. The reason for this is that I do not think that the manifestation of understanding of a phenomenon through receiving an explanation is much different for scientists than for laypeople. Laypeople also want to understand phenomena in the world, this is not an exclusive goal of scientists. And the usual way laypeople go about understanding some phenomenon they are interested in is to read literature about the respective topic or listen to talks or podcasts from specialists, which are often scientists. In such cases, the subject in question, may it be a scientist or a layperson, gets to know an explanation of a phenomenon, but has not articulated the explanation herself. In such situations, when an explanation is explicitly available already and a subject learns or receives this explanation by reading a text or by listening to an expert testimony, two different things can happen. Either, the explanation is just added to the knowledge of a subject, which is not identical to understanding the aspect of the phenomenon that the explanation represents. This is a case of simply knowing an explanation in the sense that the subject accepts, maybe even believes, the explanation, can repeat and possibly even reformulate it. Or she does grasp the relations represented by the received explanation, and through this is getting epistemic access to the phenomenon that she did not have before she received the explanation.

In this second scenario, her understanding of the phenomenon will have changed or improved, in contrast to the first scenario, in which she just gained additional knowledge of the explanation, but no understanding of the phenomenon. I elaborate on the difference between knowing an explanation and understanding the phenomenon that is represented by this explanation, through grasping, in section 4.3. However, the point I want to make here is not about the difference between knowledge and understanding, but instead about the question whether there is a difference in the understanding through receiving an explanation in cases where the subject is a scientist or a layperson. At least *prima facie*, I do not think that there is a difference. Whether one is a scientist or layperson, if one reads or listens to some explanation of some phenomenon, one will have to grasp the relations that are represented by the explanation. In contrast, achieving understanding through conducting scientific research is only possible for scientists who acquired the necessary resources and were trained to use them for the manifestation of understanding. I elaborate on these resources in section 6.2. Laypeople are not trained to be scientists, they lack the resources that scientists have, are not able to conduct scientific research and hence are not able to understand a phenomenon through this procedure. Laypeople as well as scientists are in principle able to acquire understanding of a phenomenon on the basis of an explanation they receive, but only scientists are able to achieve understanding of a phenomenon through scientific methods and an articulation of a new, previously not known or non-existent, explanation grounded in these methods.

#### 6.1.2.3 Aspects of phenomena and details of relations

Again, the objects that scientists ultimately want to understand are empirical phenomena. I already explained in section 6.1.2.1 that relations of phenomena need to be grasped for understanding, as understanding is something like "seeing how things hang together". However, I introduce additional qualifications in the *manifestation condition* of understanding, namely that *details of* relations of *aspects* of *P* should be grasped and explained. Why do I introduce these additional restrictions? Because, at least in the vast majority of cases, phenomena are not fully understood in the course of a single study. In other words, (most) phenomena that are of scientific interest are so complex that it is impossible to grasp and explain all the relations a phenomenon stands in at once, or even to grasp all the details of one relation between only two aspects of a phenomenon at once, as individual relations may also be quite complex. Usually, scientists perform several experiments, compare many samples, and collect a lot of data to discover aspects of a phenomenon. Scientists understand a phenomenon in a piecemeal fashion, through grasping and explaining more and more aspects of the phenomenon, the relations between these different aspects, and also the details of any relation. This process takes time.

Consider the zebrafish episode from chapter five, where this piecemeal understanding of the target phenomenon can be seen on two different levels. On the more general level, the phenomenon that biologists working on zebrafish wanted to understand is the genetic regulation of embryonic development of vertebrates. As this is a very complex phenomenon in which various different genes interact with each other at various stages during the developmental process, the biologists had to split up in several research groups, focusing on specific genes in their respective laboratories. Some groups restricted their research to the development of the cardio-vascular system, others to the nervous system, etc. In other words, the various research groups grasped (and subsequently explained) different relations between some aspects of embryonic development, namely those they were researching. The results gained in the different zebrafish laboratories are shared with the community, for example via the Zebrafish Information Network (ZFIN), an online database, to enable colleagues to use results for their research on a different aspect of embryonic development. On the level of a particular research group, even the understanding of the function of one specific gene takes place step by step. Recall the research on the *oep* gene that I discuss at length in chapter five. The results on the function of *oep* for the Nodal signalling pathway, which were published in one single article, required several experiments, each one focussing on one specific feature of the function of *oep*. For instance, one experiment was required to test whether Oep is necessary for Nodal signalling or whether it has merely an amplifying function, while another experiment was necessary to discover the exact location of Oep in the Nodal signalling pathway. Again, these are just two out of more experiments that one group of scientists conducted on *oep* in zebrafish.[12]

Hence, it is neither necessary nor always the case that a phenomenon is understood in merely one manifestation process of grasping and explaining. On the contrary, it only rarely is the case that a phenomenon is fully understood by a scientist through one manifestation process. Furthermore, various details of a relation or several relations that are relevant for understanding (aspects of) a phenomenon may be grasped successively, especially when the investigated phenomenon or the relations involved get more complex. It happens that first the presence of a relation is grasped, and that further details of this relation are worked out in the course of further investigation. It may even be the case that some details of a relation become epistemically accessible only after other details of the relation or further relations were already grasped and articulated in hypothetical explanation. This feature can be observed in the episode of the research on *oep*, too. The goal of the scientists in this episode was to understand the function of the *oep* gene in embryonic development. When they generated zebrafish mutants that lack the *oep* gene, the process of understanding started by first grasping the similarity relation between the phenotypes of the generated *oep* mutant and the *sqt/cyc* double mutant. This is a relation between the phenotypes of two mutants. Based on the knowledge that the *oep* mutants lack the Oep protein, the *sqt/cyc* mutants lack the protein Nodal, and on the observation that both mutant strains have a similar phenotype, the biologists concluded that the proteins Oep and Nodal must be related somehow. Only due to the grasping of the similarity relation between the *oep*- and the other mutants was it possible for the biologists to grasp, to get epistemic access to, a relation between the proteins Oep and Nodal and, therefore, also between the respective genes. Only now had the biologists *reasons* to assume that there is a connection, but they did not understand this connection, yet. To gain understanding of what is going on and how the genes involved interact, the biologists investigated the observed similarity of the different mutants further and considered reasons for it. They were looking for an explanation of the similarity and had the idea that Oep and Nodal might act in a common pathway. If this is the case, it would explain why the same phenotypic effects can be observed when one of the two components of the same pathway is missing. In both cases, the pathway would not function properly and lead to identical effects.

---

12 See Gritsman et al. (1999) for information concerning all the experiments in this study.

This was the first step in the manifestation of understanding of the function of *oep* in embryonic development. The biologists understood that *oep* is somehow related to *nodal* (or its orthologs). However, they had no idea what this relation may look like, or how exactly these proteins interact. They had not grasped the details of the relation between *oep* and *nodal*, yet. In a second step, the functioning research infrastructure came into play. After arriving at the idea that Oep and Nodal might somehow act in a common pathway, the biologists referred to the results and knowledge about the proteins, and also the genes coding for the respective proteins, previously gained in other studies, not in their own. Research on Oep as well as on Nodal in zebrafish was already conducted, but independently from each other. A relation between the respective genes had not been assumed before. The integration of additional knowledge "suggested that Oep is required for cells to receive Nodal signals."[13] This is already a much more concrete conception of the relation of *oep* and *nodal*, more concrete than the insight that the two genes are somehow related. The biologists arrived at the hypothetical explanation that *oep* has an important function in vertebrate embryogenesis, because it activates the Nodal signaling pathway by which germ layer formation, organizer development, and the positioning of the anterior-posterior axis are regulated.

However, this was only a hypothesis or a hypothetical explanation. The scientists wanted to have supporting evidence to ensure that they understood the function of *oep* correctly. In a third step, the biologists designed and conducted several experiments to determine whether Oep is indeed necessary for Nodal signaling and where in the pathway Oep is located exactly. For example, through counterfactual reasoning the scientists came up with an experiment to test the counterfactual situation that Oep is *not* necessary for Nodal signaling. This experiment did decisively show that Oep is indeed essential for Nodal signaling, and not merely an amplifier. The biologists could confirm their hypothetical explanation, that they actually grasped a detail of the relation in question. Nonetheless, before conducting the additional experiments to investigate aspects of the function of Oep (the phenomenon) in the third step of the manifestation of understanding, the biologists could not know whether their articulated hypothetical explanation is correct. That is, they could not know whether they already understood the function of *oep*, that they actually grasped a detail of the relation between *oep* and *nodal*, or rather misunderstood it. The experiment in which the biologists tested the need of Oep for Nodal signaling could have falsified the hypothetical explanation and the biologists would have realized that they had misunderstood the function of *oep*, i.e., that their explanation did not match the phenomenon. In this episode, the articulated hypothetical explanation was confirmed, but it does happen that hypothetical explanations are falsified, which indicates that researchers misunderstood the relations of the phenomenon

13 Ibid. p. 125.

in question. When phenomenon and explanation are conflicting with each other, the conflict (usually) motivates scientists to articulate an alternative explanation, to understand the phenomenon in a different way in which the conflict dissolves.

In sum, in order to understand the function of *oep* in embryonic development (the phenomenon), the scientists successively grasped and explained the similarity relation between the *oep*- and other mutants, then the relation between *oep* and *nodal* (or its orthologs), and subsequently the *details of the relation* between *oep* and *nodal*, e.g. that *oep* is necessary for activating the Nodal signaling pathway and not an amplifier, as well as the exact location of Oep in the pathway, i.e., the receptor and transcription factor with which Oep interacts. Hence, the *parts or aspects of the phenomenon* between which these relations hold include the phenotypes of the mutants, the genes and respective proteins. It is not sufficient for understanding to merely grasp the parts or aspects of the phenomenon independently from each other, as this would not allow to "see how things hang together". Rather, one must grasp which aspects or parts are related, and how exactly they are related. Or consider the famous flagpole example. If you want to understand why the shadow of a flagpole has a certain length, it will not be sufficient to grasp or recognize the length of the shadow, the length of the flagpole and the position of the sun (aspects of the phenomenon) without grasping how these aspects are related.

#### 6.1.2.4 The gradual nature of understanding and the iterative nature of its manifestation

So, as scientists understand the phenomena they are investigating step by step, the understanding of the respective phenomena manifests in the successive grasping and explaining of details of relations that hold between different aspects of the phenomena. The manifestation process of scientific understanding is schematically shown in figure 2.

The iterative nature of the *manifestation condition* accords with one feature of understanding that is so far uncontroversial in the philosophical debate on understanding, namely its gradual nature. Understanding comes in degrees, as I already said in section 4.2.2. The GE-account adheres to the procedural nature of understanding and accommodates it by an iterative, stepwise process of grasping and explaining (details of) relations of (aspects of) phenomena or several relations of several aspects one after another. The understanding is complete for a specific scientific episode if all relevant relations and their details are grasped and articulated in explanations. Following every instance of grasping, the scientists reason or reflect about the grasped information, articulate them in a hypothetical explanation and continue to explore the phenomenon if they identify further questions concerning the phenomenon they cannot answer yet.

*Figure 2: The manifestation of scientific understanding.*

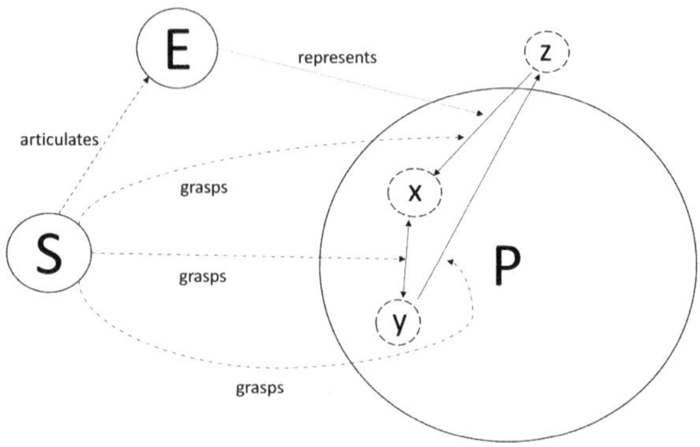


*S grasps relations $R_1$, $R_2$, and $R_3$, which hold between the aspects x, y, and z, where z is some external aspect that is related to the internal aspect x, and articulates a new explanation $E_{R3}$ that represents $R_3$. The details of $R_3$, which are not depicted here, would include, for instance, what kind of relation it is. If $R_3$ turns out to be a causal relation, a further detail will be that z causes x.*

If scientists find gaps in their newly acquired knowledge of a phenomenon, which they cannot close by referring to available knowledge, they will conclude that their understanding is incomplete. That is, they have not grasped every detail of a relation, or all relations involved, or could not articulate all the grasped information, yet. Importantly, notice that I use the notion of 'complete understanding' in reference to a *specific research episode*. I do not talk about complete understanding as a context-independent, ideal understanding. Rather, I view understanding to be complete if scientists within a specific research episode answer all research question about a phenomenon they wanted to answer in this episode. This does not exclude the possibility of investigating the very same phenomenon again in the future and asking new or possibly even the same research questions in light of new knowledge or evidence again. However, in order to gain any scientific understanding, specific resources are required.

### 6.1.3 The *resource condition*

Therefore, I call the second condition of the GE-account *resource condition*, as it captures all the resources that scientists need for understanding a phenomenon through scientific research. The analysis of the scientific episode from biology in chapter five revealed the necessity of these resources, which cover (material) equipment, relevant knowledge and research skills. In this section, I argue that these resources are, in an abstract sense, necessary for any kind of scientific research in any scientific discipline, and hence also for scientific understanding.

#### 6.1.3.1 (Material) Equipment

The insight that scientists need adequate equipment to do research at all should not come as a surprise. Neither does the observation that not every phenomenon can be investigated with any equipment. A microscope will not enable anyone to observe stars and other planets. While the awareness that the research on different phenomena requires diverse equipment will probably be seen as trivial for many readers, I think it is important for an account of scientific understanding to mention this insight explicitly. This is so for the mere fact that the existence and availability of equipment has a direct and grave impact on the possible understanding that scientists could acquire of any phenomenon. Leaving this fact unmentioned would not do justice to the fundamental influence of the equipment on understanding.

First and foremost, one cannot try to understand a phenomenon one is not aware of. For instance, in order to understand global climate change, one must first realize that something like global climate change exists or takes place. To achieve this, certain equipment like thermometers are already required in a sufficient quantity. Once some phenomenon is discovered, scientists need to engage further with it in some way to understand it. For this further engagement, additional and potentially divers equipment is necessary. In order to understand the mechanisms involved in or driving climate change, scientists need, for example and among other things, appropriate computer models and computing capacity to run their simulations.

Two remarks are in need. First, throughout chapter five, I used and referred to the term 'tool'. I do not use this term here, as I subsume tools under the notion of equipment. For doing research, scientists need tools as well as the stuff or material to apply the tools to. Thermometers as well as computer simulations can be viewed as tools, and hence belong to the equipment of climate scientists. Second, computer models and the data that are used by scientists, not only in climate science, are not strictly speaking 'material', while computers on which the simulations are run and the data stored definitely are material objects. This is the reason why I bracket the attribute 'material' when talking about the equipment. Without some material equipment like computers or hard drives, at least, no research will be possible, including disciplines like theoretical physics or theoretical chemistry that, at first glance, do not use 'material' equipment in their investigations. Hence, in theoretical disciplines like theoretical physics or chemistry, or even in the humanities, the (material) equipment may play a subordinate role in the acquisition of understanding. However, some minimal (material) equipment like books, writing material or computers will be necessary in such theoretical disciplines. Furthermore, many of the empirical sciences require much more and divers material equipment than theoretical disciplines. In the scientific episode on understanding the genetic regulation of vertebrate development presented in chapter five, the list of necessary equipment includes an appropriate model organism, or even a specific mutant strain of that organism, depending on the research questions asked, adequate aquaria systems, dissection microscopes, devices to induce mutations, for instance through γ-radiation, chemicals or induction of mRNA, and descriptive devices including cellular fate maps, neural wiring diagrams, or staging series. Again, which equipment is required precisely is dependent on and needs to be analyzed for the concrete scientific episode, given the respective research discipline and the phenomenon that is investigated.

In short, (material) equipment is necessary for conducting scientific research on phenomena, and hence for gaining scientific understanding of these phenomena. While this insight may not be very novel, it should be made explicit, as the available equipment determines one contextual dimension that impacts the possibility of understanding. Depending on the available equipment, it is possible to understand some phenomena, but impossible to understand others. The existence and use of specific (material) equipment makes understanding possible in the first place. Biologists would not have been able to acquire understanding of the genetic regulation of vertebrate development if they had not introduced zebrafish as a model organism to work on and created and assembled additional equipment, like the devices to induce mutations and to identify phenotypic effects. Yet, the appropriate (material) equipment does not exhaust the resources needed for understanding. Another resource that scientists need as well is relevant knowledge.

#### 6.1.3.2 Relevant knowledge

I take knowledge to be necessary for scientific understanding, in the sense that a scientist cannot understand a phenomenon if she does not know anything about it. I consider knowledge to be propositional and I subsume concepts like natural laws, theories or empirical data under the term knowledge. If a scientist wants to understand a certain phenomenon, she must start somewhere and must draw on theoretical and empirical background knowledge that has been established and accepted by the scientific community the scientist belongs to. This claim accords with Michael Polanyi's analysis of the interconnectedness of articulate and inarticulate intelligence, which I present in more detail in sections 4.2.3 and 4.3, and his conclusion that humans cannot understand the world without reference to an established articulated conceptual framework. Polanyi argues that humans always rely on the knowledge about the world that previous generations collected and stored in the respective language of a community. This is the case for every human community, not only different scientific communities. Resorting to knowledge about the world already established and stored in language in the past enables new generations to directly approach new problems, questions, or phenomena, instead of starting from the beginning all over again. In the case of science, students acquire the necessary knowledge in lectures and seminars during their studies. For example, physics students learn the basics in mechanics, optics, electromagnetism or solid state physics, together with the respective vocabulary that the physical community developed to represent and store knowledge in these fields. Already established knowledge is taught to junior researchers, in order to enable them to make use of that knowledge when addressing new phenomena and unanswered questions in all scientific disciplines, not only in physics, of course. Without taking some knowledge as an established basis, no new knowledge could ever be gained, no progress in scientific knowledge could be made.

The knowledge that is already established and available in a concrete research episode is another contextual factor that influences the understanding that can possibly be acquired of a phenomenon. Depending on what scientists know and do not know (yet), they may be able to understand some phenomena, but not others. Take again the episode on zebrafish from chapter five. Molecular geneticists wanted to understand the effects of genetic interactions on the development of vertebrates, but the knowledge that they possessed within their discipline was insufficient to understand this particular, though complex, phenomenon. Molecular geneticists had only been concerned with molecular processes within a cell, and their available background knowledge enabled them to address phenomena in this domain, but they could not exceed it. For addressing and understanding developmental phenomena, knowledge about molecular features or processes was not enough. Additionally, knowledge about cell, tissue and organ properties as well as organism as wholes was required. And molecular geneticists acquired this knowledge through cooperating with developmental biologists who possessed it and were interested in the same phenomenon as the molecular geneticists. As I elaborate in section 5.2.1, developmental biologists had the same problem as the molecular geneticists at the beginning of the research episode, namely that they were lacking necessary knowledge. The developmental biologists did not know anything about molecular mechanisms or properties or about genetics. Since the background knowledge from molecular genetics and developmental biology complemented each other, and researcher from both disciplines had a shared interest, the cooperation and, ultimately, integration of the disciplines was fruitful for the understanding of the genetic regulation of vertebrate development.

Notice that I do not claim that an integration of different scientific disciplines is necessary for or always a guarantee for achieving understanding for some phenomenon. It happened that in the episode from scientific practice that I have chosen that an integration of two research disciplines fruitfully enabled understanding, but this may not always be the case. Attempts of integration or even merely cooperation of different disciplines may also fail, and in many cases scientists within one disciplines are able to generate new knowledge that they need on their own, without interacting with any other discipline. In short, I am not making any claims about the role of integration of various scientific disciplines for scientific understanding. Such a form of integration may fruitfully enable or foster understanding of certain phenomena, as in the research on zebrafish, but it does not have to, necessarily. The only claim I am making here is that for understanding a specific phenomenon, specific knowledge is required. One cannot understand the genetic regulation of vertebrate development without having knowledge from molecular genetics as well as from developmental biology, one cannot understand global climate change without some knowledge from physics, one cannot understand potential effects of a high inflation without some knowledge from economics. How scientists acquire the knowledge that they need for understanding a specific phenomenon, whether they generate this knowledge themselves within their own disciplines before addressing the respective phenomenon, cooperate with scientists from another field or establish a new discipline through an integration of several already existing disciplines, varies depending on the episode one looks at.

So, with the term knowledge I am referring to every kind of propositional knowledge that may be relevant for the phenomenon under investigation and already contained in the background knowledge of the scientist or in the informational sources of the research community. All scientists rely on the already established background knowledge of their community when conducting their research, answering new research questions, and generating new knowledge. Knowledge must be explicit or made explicit when necessary.[14] Whether a scientist can understand a specific phenomenon depends on what she already knows or to which knowledge she has access and on the available (material) equipment she could use in her investigations. Yet, another type of resource necessary for understanding is missing.

#### 6.1.3.3 Research skills

In addition to the equipment and knowledge, various research skills play a necessary role for scientific understanding, too. In contrast to my notion of knowledge, I view skills to be non-propositional. The concept of skills or abilities is discussed at length in chapter four, where I develop and apply the following definition of an ability:

14 Again, I subsume theories under my notion of knowledge and will not address more specific possible functions of theories for (scientific) understanding. In this respect, the GE- account differs from Henk de Regt's account, in which scientific understanding of phenomena can only be acquired on the basis of theories. I discuss the reasons for and the advantages of not giving theories a special status in the GE-account in section 6.2.1, where I compare the GE-account of scientific understanding with the account of Henk de Regt.

*x* is an ability if and only if *x*

i) is a disposition to perform a cognitive or physical activity successfully with respect to relevant standards,
ii) has been learned and trained in a specific social context, and
iii) manifests in processes that are partially tacit (i.e., that can never be made fully explicit).

Remember that the terms 'ability', 'skill', and 'know-how' are often used interchangeably. I think that there is only a terminological difference between these notions. This is because expressions like 'someone has the ability to x, has the skill to x, or has the know-how to x' all amount to the same thing in the end. They all denote that someone can do something in an appreciated or valued manner. Hence, understanding as well as research skills, which are the topic of this sub-section, fall under my definition of ability. However, understanding is a different ability than the abilities I subsume under the term research skills. Understanding is the ability to make sense of a phenomenon in a scientific way, while research skills are abilities needed for conducting scientific research, e.g. taking measurements with specific devices, collecting samples, or programming computer simulations. I view understanding to be a more holistic ability than the research skills needed to gain understanding. So, for the sake of clarity, I refer to understanding as an ability and to all other kinds of 'know-how' that contribute to understanding in the scientific context, the research skills, as skills.

Research skills enter the scene in the play of understanding by actually using the available equipment and knowledge in order to really do scientific research. Research skills cover all skills that scientists learn and employ in the scientific practice of their discipline. These research skills are required to set up an environment in which new (hitherto unknown) relations can be grasped (i.e. in which a phenomenon can be investigated) and an explanation based on investigating the grasped aspects of the phenomenon and on the available knowledge can be articulated. Scientists have access to information from their background knowledge and also from the current investigation of the phenomenon, for example through observations, measurements, or modelling procedures. This information has to be selected, used, and reasonably connected to grasp relations and articulate an explanation of the phenomenon. There are no fixed rules how exactly this should be done.[15] Depending on the object of understanding, the training of the scientist, which information are

15 De Regt argues for this characteristic of scientific understanding as well, see de Regt (2017), chapter 2.2.

available, how they are (re-) presented and the epistemic goals of the scientist, it varies significantly which relations are grasped by scientists and how the resulting explanation looks like. This is due to the observation that understanding is a multi-track disposition, a feature for which I argue in section 4.2. Depending on their discipline and the historical context, scientists not only have different bodies of information (i.e., knowledge) and varying equipment at their disposal, but they also learn different research skills, e.g., handling a particular measurement device in the laboratory, or using varying modelling or statistical tools. The research skills scientists learn and apply shape the acquired understanding. Research skills, together with the available knowledge and equipment, have an impact on which relations can possibly be grasped and which information are put in the explanandum and in the explanans, that is, which pieces of knowledge are associated with which aspects of the phenomenon and how they are connected (i.e., what kind of relation holds between knowledge and phenomenon or between aspects of phenomenon, i.e., causal, deductive, probabilistic, mechanical, functional,... relations).

The necessity of research skills to understand a particular phenomenon has also been highlighted in the episode on the zebrafish research in chapter five. In order to understand the genetic regulation of embryonic development of complex organisms like vertebrates, sophisticated research skills were as much required as appropriate equipment and relevant knowledge. As for gaining the required knowledge, the integration of molecular biology and developmental biology also provided the possibility for the involved scientists to learn and practice the research skills necessary for the envisioned research. The molecular biologists had the skills to induce and map genetic mutations in zebrafish, but were not able to relate the insights gained through this procedure to any effects that the mutations have on the phenotype of the embryos. In fact, they were neither able to identify any phenotypic effect, nor to actually do research on biological structures that exceed the molecular level. Molecular biologists had the research skills to engage with molecular mechanism, but they never acquired the research skills to work with more complex tissues, organs, or even embryos as a whole. And the developmental biologists, in contrast, had the research skills to identify and work with phenotypic effects, they were able to dissect embryos, but were not able to engage with phenomena on the molecular level through, for example, mutational analysis, because they had never learned to do mutational analysis. Because of the lack of specific research skills necessary for the phenomenon in question on both sides, none of the scientists involved in the early stage of the new research endeavor could investigate the phenomenon they wanted to do research on, and hence no one could have grasped any of the relations involved in the phenomenon or explain anything. Only through the integration and acquisition of research skills from both biological disciplines were the scientists able to do the imagined research, to grasp relations between a mutation and its phenotypic effects in the development of an embryo, and ultimately to draw conclusions on genetic interactions in normal developmental processes. In a nutshell, without the implementation of the respective research skill, the research on zebrafish could not have been conducted, not to be mentioned that biologists would have understood anything.

So, while possessing the required research skills is necessary for grasping relations and articulating explanations, it does not automatically amount to grasping relations and articulating explanations. Therefore, I distinguish between research skills on the one hand and grasping and explaining, alias the manifestation process of understanding, on the other. It can still happen that a research project, despite being conducted properly, does not provide the insights scientists expected. The empirical data that are obtained in a research project may not allow for grasping any hitherto unknown relation, despite the fact that the involved scientists used their research skills appropriately. There is no guaranty that any study, or the data obtained by it, provides new insights into the phenomenon, that it reveals new aspects so that scientists could grasp them. Not every study enables epistemic access to a (so far) hidden aspect of the phenomenon. And even if a scientist is able to grasp a hitherto unknown relation of the phenomenon on the basis of her background knowledge and the appropriate application of research skills, the articulation of the grasped relation in form of an explanation, and hence the understanding of the respective aspect of a phenomenon, may still not be possible. This was the case with James Clerk Maxwell and his attempt to understand the specific heat anomaly of gases like oxygen or nitrogen, an example that I already briefly referred to in section 4.2 and 4.3 and which has been analyzed in detail by Henk de Regt.[16] Maxwell failed to understand the specific heat anomaly, because he could not articulate an explanation of why the anomalous gases have the specific heat ratios that were determined empirically. He got a grasp on the relation between the specific heat ratios and the kinds of molecular motion that these gases exhibit, and introduced the concept of degrees of freedom in his attempt to explain the specific heat ratios based on the kinds of motion of the gas molecules. Still, the explanans he articulated did not accommodate the explanandum, the empirically determined values of the specific heat ratios of the anomalous gases. Hence, Maxwell failed to come up with an explanation of the specific heat anomaly. This phenomenon remained a mystery to Maxwell, he did not understand it, despite the fact that he possessed impressive research skills and was one of the most outstanding physicists in the nineteenth century. It was Ludwig Boltzmann who used the concept degrees of freedom, introduced by Maxwell, to develop his dumbbell model of the anomalous gases, who provided an explanation of the specific heat ratios based on this model, and hence understood this phenomenon.

16 For an in-depth analysis and discussion of this episode from scientific practice, see de Regt (2017), pp. 205–216.

Research skills are intertwined in the manifestation process of understanding, to first grasp and then, subsequently, articulate new discoveries relating to the phenomenon in an explanation. Understanding is the ability to make sense of new discoveries about a phenomenon in the context of already available knowledge. Research skills are required prior to grasping any relations, for setting up experiments, conducting measurements, or for analysing data appropriately so that it becomes possible to grasp, to get epistemic access to, some relation of the phenomenon in the first place. The biologists in the episode on zebrafish would not have been able to grasp a relation between Oep and Nodal if they did not had the research skills to, among other things, generate the Oep-mutants through cloning techniques and identifying phenotypic effects of these mutants. And also after some relation is grasped, research skills are required again to investigate that relation and to arrive at an explanation. For instance, after a relation between Oep and Nodal was grasped, it should be clarified whether Oep has an activating or amplifying function in the Nodal signalling pathway. Hence, the biologists needed the research skills to set up experiments in which they could test precisely these possibilities. If they lacked these research skills, they would not have been able to discover that Oep has an activating function. They would not have been able to explain the function of Oep for Nodal signalling and, therefore, would not have gained understanding of this phenomenon without the research skills to set up and conduct these experimental studies. Understanding manifests in an iterative process, as I argue in section 5.2.3, and for every iterative step of grasping and explaining, scientists need specific resources.

Scientific understanding requires research skills, knowledge, and specific equipment. These three kinds of resources are means that serve the end of understanding. If a scientist is lacking any of these necessary resources, she will not be able to grasp or explain relations of a phenomenon, since she would not be able to research that phenomenon at all. Knowledge, research skills, and also material equipment are necessary for understanding, because their availability and application *enable* scientists to do research, as well as to grasp relations and to articulate them in form of an explanation. In the episode on zebrafish, these were relations between genetic activities and developed phenotypes and also epistatic interactions among genes. It would not have been sufficient if the biologists only had all the propositional knowledge from molecular and developmental biology and all the necessary equipment, including zebrafish mutant strains, aquaria systems, dissection microscopes etc., but would have lacked the research skills. They would not have been able to apply the knowledge and equipment to really carry out an experimental study. Merely possessing theoretical knowledge does not allow for relating this knowledge to phenomena in the world, and thereby understanding them. The same applies for merely possessing the necessary equipment. Simply having the required material and instruments is not identical to having the skills to use them. And being well trained in all the required research skills also only enables understanding if the required equipment and relevant knowledge is available, too. Having the skill to use a dissection microscope, for instance, will not be of any usage for a scientist if she does not also have a dissection microscope and some organism to dissect. And if a scientist can accurately manipulate genes, dissect embryos, and apply statistical tools, if she has the research skills and equipment, she will not be able to recognize any significant effect if she does not know what to look for, or will not be able to make sense of anything that she may recognize if she does not also possess the, in this case, relevant knowledge. Again, this is so because humans cannot understand anything in the world without relying on some already established background knowledge.

#### 6.1.3.4 Having all the resources

So, this second condition of the GE-account, the *resource condition*, explicates which resources need to be available for scientists so that they can possibly understand a phenomenon through research. These resources are (material) equipment, relevant knowledge, and research skills, as figure 3 illustrates.

*Figure 3: The resources for scientific understanding.*

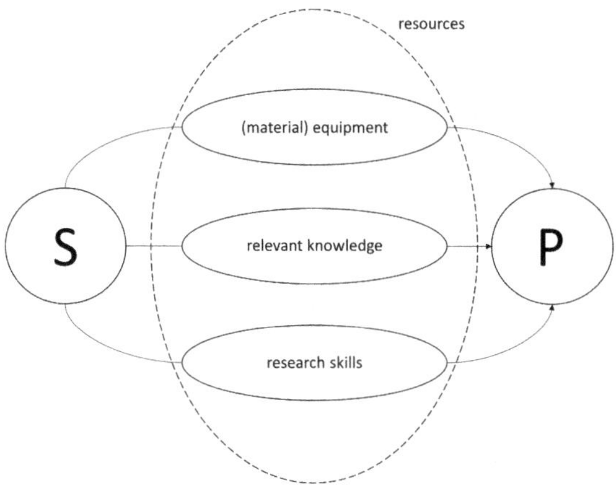


*S needs the resources to do research and scientifically understand P.*

What kind of equipment, which knowledge and research skills are needed to understand some specific phenomenon depends on the phenomenon and has to be analyzed in the individual cases. Importantly, the availability of these resources is necessary for understanding a phenomenon, it enables understanding, but it is not sufficient. Even if a scientist has all the resources at her disposal, she still has to manifest the ability to understand the phenomenon. She still has to grasp relations and articulating explanations, and hence to fulfill the *manifestation condition*. Yet, one final condition is still missing, namely the *justification condition*.

### 6.1.4 The *justification condition*

Scientific understanding of a phenomenon is context-sensitive and hence influenced by historical, disciplinary, and social factors, as I already alluded to in section 6.1.1. In other words, the scientific community in which an individual researcher is embedded impacts the understanding that she can achieve. More precisely, the scientific community serves two functions regarding the understanding of individuals. First, the community has to enable its members to gain scientific understanding of phenomena. And second, parts of the community also have to assess and approve the understanding that its members achieve as scientific. Let's have a look at these two functions, which resemble the famous distinction between the context of discovery and the context of justification.

#### 6.1.4.1 Enabling a scientist to understand a phenomenon

Most basically, every first semester science student joins the scientific community as a whole. Of course, *the* scientific community, referring to the sum of all scientists, is a fairly vague ascription and can be split up into sub-communities along various dimensions. One of these dimensions is the respective discipline, like physics, chemistry, biology, psychology, geology, and so on. Scientific communities belonging to these disciplines can be subdivided even further. Within biology, for instance, we have genetics, physiology, botany, zoology, ecology and many more, and these subdisciplines again cover several sub-communities that are even more specialized in some way. And even within one and the same (sub-) discipline, there is historical variation. Every discipline changes and develops in some way in the course of its history, for example by changing its methods and scientific standards. And it also happens that two or more disciplines merge in order to cope with new phenomena, as it happened in the episode on zebrafish, in which researchers from molecular genetics and developmental biology founded the new discipline of developmental genetics. I do not want to argue for any specific conception or definition of what a scientific community is or may be. Maybe there is no single and strict definition of a scientific community, as communities themselves change in the course of history. Fortunately, this is not a problem for the GE-account of scientific understanding. Every young science student becomes a member of some scientific community, whatever its demarcation to other disciplines or communities may be and whether this demarcation is fluid or not. Throughout their careers, scientists get more and more specialized within their discipline, within their community. To put it differently, scientists specialize in a way that is necessary to gain understanding of the phenomena they are interested in and that are addressed by the specific community they joined. If necessary, this specialization includes a broadening of the discipline, inter- or transdisciplinary research, the collaboration with scientists from other communities or even an integration of several disciplines, as in the zebrafish episode.

Within their scientific community, researchers (ideally) get all the resources they need for doing research on and to understand phenomena. These resources include (material) equipment, already established knowledge, and research skills, as is already explained in section 6.1.3. But furthermore, science students also learn and practice the ability to understand phenomena with which their community is concerned scientifically. Recall that I argue in section 4.2.3 that, in general, understanding is the ability to make sense of an object (a situation, an experience, or a phenomenon) by aligning the object with the language used. In the case of scientific understanding, young scientists acquire the ability to make sense of some phenomenon in a way that is accepted by parts of their community as scientific. They learn to grasp relations that are relevant for the phenomenon in question and articulate these relations in explanation by using adequate background knowledge through exercises or tasks provided by their professors and supervisors. In the course of lectures, seminars, laboratory courses of field trips, supervisors show how open questions or problems concerning some phenomenon are addressed and solved in the respective discipline, how scientists in the discipline understand phenomena they are researching. And then, students or young scientists are confronted with exercises they have to solve themselves. They have to demonstrate that they are able to make sense of, to understand, phenomena on their own. This description of how young scientists acquire the ability to scientifically understand phenomena is backed up by my discussion of how any ability, not only understanding, is learned, which I present in section 4.1.3. There, I argue that any ability can only be learned by practice within a community and guided by a master, teacher, or supervisor, and not from a textbook.

The various crucial functions of the scientific or disciplinary community for scientific understanding and how understanding is contextually influenced becomes apparent in the episode on the research on zebrafish as well. Scientific understand of the genetic regulation of vertebrate development, as in the case of the *oep*-gene, could have been acquired by the involved biologists only because an appropriate context and community were established. Only through the integration of molecular and developmental biology could the researchers on zebrafish acquire the knowledge and train the research skills from both disciplines which they necessarily needed to do the research they wanted to do. Additionally, through conducting *The Big Screen* in the second stage of the episode, the zebrafish community provided itself with the required material resources such as the zebrafish mutant strains and the laboratory equipment, like sophisticated aquaria systems. Having all the necessary knowledge and research skills will not provide understanding if a scientist does not have the material equipment to work with. The availability of all the mutant strains enabled research that would not have been possible prior to *The Big Screen*. And the researches working on *oep* did not only need the respective mutants, but also access to knowledge generated by other research teams to make sense of the similarity they observed, to come up with ways to investigate the relation between the genes in more detail, namely to grasp more aspects of the relation of *oep* and *nodal*, and to understand it at the end. Communication and exchange with other researchers is necessary in order to get all the equipment and pieces of knowledge that are required to understand a specific phenomenon in a certain context. However, the respective scientific community also fulfills a second crucial function for the understanding of individual scientists, to which I now turn.

#### 6.1.4.2 Approving the understanding of a scientist

In addition to providing any individual scientist with all the resources necessary to understand a phenomenon and to teach her how to scientifically understand phenomena, some members of the scientific community also have to assess whether she indeed did understand some phenomenon scientifically. But how can it be assessed whether a scientist has gained scientific understanding, the ability to make sense of phenomena in a scientific manner? This is possible only through an explanation that a scientist articulates and then communicates. Basically, she has to come to new knowledge by herself, not by merely reading a book or listening to someone, and she has to be able to make this new knowledge explicit. Again, the newly produced explanation or knowledge has to be new only for the reasoning subject herself, as I already explained in section 6.1.2. For assessing whether an explanation is in fact a legitimate explanation, that it accords to the existing disciplinary norms, a scientist has to communicate her new insight, the new explanation, and the way through which she arrived at that explanation, to other scientists. Communicating the grasped relation in form of an explanation and the methods and practices through which one arrived at that explanation is the only way for other members of the community to assess whether an individual has in fact understood the phenomenon in question in an appropriate manner.

It is necessary for scientific understanding that scientists articulate explanations, because this is the only way that scientists can make their understanding of a phenomenon explicitly and publicly accessible. Understanding is a cognitive ability and its manifestation, the grasping of relations and articulation of explanation, is a cognitive process. This process is hidden from other members of a research community since scientists have not (yet) found a way to peek into the head of their colleagues and see their thoughts or inferences they make. Only the result or product of the manifestation of understanding, an explanation, can be presented and is, hence, accessible to other members of the community. And scientists should want to make their understanding assessable by making an explanation explicit, because they want to get things right. Ultimately, scientists want to discover, know and understand how the world really is. They aim at having justified beliefs and avoid a reliance on luck, they strive to get the best possible confirmation and justification that their understanding is correct (in terms of the contextual standards of the discipline). To achieve this, scientific understanding has to be made accessible for colleagues.

This idea of seeking confirmation and justification for the individual scientific understanding and thereby increasing its objectivity by appealing to the scientific community is in line with the views of Helen Longino and Heather Douglas, who take objectivity (of hypothesis, explanations, theories) to be a feature of a social community. While both reject a strong notion of objectivity in terms of the value-free ideal, they argue for a conception of objectivity in terms of intersubjectivity reached through social processes like critical discussions.[17] Although neither Longino nor Douglas were explicitly concerned with scientific understanding, de Regt argues that their analyses of objectivity can also be applied to understanding.[18] That is, "whether or not the understanding that is produced may be considered objective depends on whether the individual and social processes conform to the given conditions of objectivity."[19] The notion 'objective' in this quote refers to my usage to the term 'scientific'. That is to say, if the understanding that some scientists achieve is labelled scientific, this understanding will be regarded as objective by parts of the respective community. And Catherine Elgin holds that a scientist, in her role as an epistemic agent within an epistemic community, has to stand "not just in a suitable relation to the phenomenon she seeks to know or understand, but also in a suitable relation to other members of the epistemic community"[20] and elaborates "the obligations that members of the scientific community bear to one another, and [how] these obligations infuse the epistemic goals of science."[21] That is, no scientist can be sure that she did understand a phenomenon scientifically, that her understanding is in line with the epistemic standards of her discipline, without some members of that scientific community accepting the articulated explanation, and hence the ability to understand which was manifested in that instance, as legitimate.

---

17 See Longino, H. E. (1990), *Science as Social Knowledge: Values and Objectivity in Scientific Inquiry*. Princeton, (NJ), Princeton University Press; and Douglas, H. E. (2004), "The Irreducible Complexity of Objectivity." *Synthese*, 138, pp. 453–473, DOI: 10.1023/B:SYNT.0000016451.18182.91.

18 See de Regt (2017), pp. 41–44. A further discussion of the relations between individual and collective levels of understanding can be found ibid. pp. 88–91.

19 Ibid. p. 43.

20 Elgin (2017), p. 121.

21 Ibid. p. 149. See ibid. chapters 5 and 6 for a full discussion.

In sum, the criteria for assessing whether the understanding of a scientist counts as scientific understanding depends on the scientific community, and are therefore subject to historical and disciplinary variation. Of course, all of this does not ultimately ensure that whole or parts of scientific communities will not be led astray in their understanding of phenomena. There have been cases throughout history in which some scientist understood a phenomenon in a way that was not accepted by (parts of) her community as being legitimate, while some years or decades later it turned out that the scientist in question actually got things right. Any (parts of a) scientific community at any point in time can be wrong in the assessment of the understanding that some of its members gained. As long as (a part of) the community provides good, legitimate reasons why it does not accept some understanding, this is not a flaw, as it may be the case that the understanding really is illegitimate in some specific context. As this context changes during time and along other dimensions, the assessment of someone's understanding may change with the context. So ultimately, this feature fits into the context-sensitive nature of understanding. The context does not only influence whether understanding of some phenomenon is possible at all, but also which understanding of the phenomenon is legitimate or scientific.

#### 6.1.4.3 Being justified in one's understanding of a phenomenon

Every scientific discipline is a community endeavor. Scientists work in groups or teams, they meet and discuss their projects at conferences, workshops or during lunch breaks, they rely on the research and results from their former and current colleagues, and they distribute resources. Science as we know it today is not pursued by an individual in isolation. It would not be possible to conduct science without being a member of a scientific community, because one would not have access to the required resources one needs to perform any kind of research, and to acquire the ability to understand phenomena scientifically, through conducting scientific research. So, the scientific community is important for the individual scientific understanding in two respects: first, by providing the individual with all the available resources (knowledge, research skills, material equipment) and training her understanding, the community makes it possible for an individual to grasp relations of the phenomenon, to get access to a phenomenon, and to articulate explanations in the first place. One may say, the community is crucial in the context of discovery. And second, after an individual gained some understanding of the phenomenon she was researching, she presents the results of her understanding in form of a potential explanation to parts of her community to gain additional justification that her understanding is probably correct in light of the available evidence and the upheld standards. The presented explanation may be accepted immediately, or reviewers might demand more experiments, more data, or a re-articulation of the proposed explanation until it gets accepted. Therefore, the scientific community also plays a crucial function in the context of justification of scientific understanding. Figure 4 illustrates these two functions of the scientific community for the individual's understanding.

*Figure 4: The function of the scientific community for the understanding of S.*

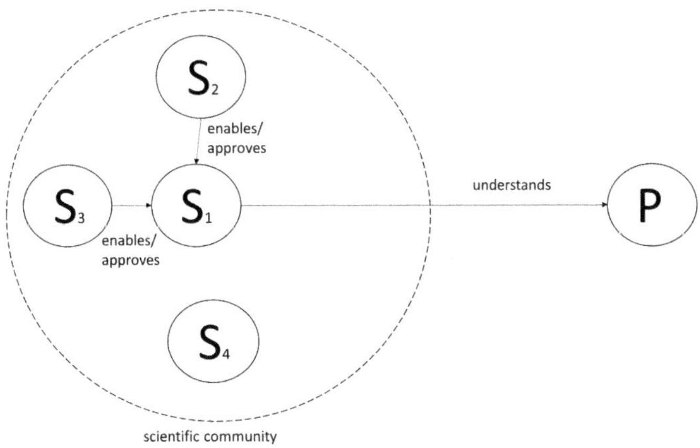


*$S_1$ understands P, $S_2$ and $S_3$ enable as well as approve $S_1$'s understanding, while $S_4$, who also belongs to the scientific community, might not be aware of $S_1$.*

### 6.1.5 Understanding – a complex ability

In this section, I presented and argued for the GE-account of scientific understanding, according to which understanding is an ability possessed by individual scientists. It is the ability to make sense of a phenomenon through research, by which new explanatory knowledge is produced. The ability to understand a phenomenon manifests in the processes of grasping relations the phenomenon stands in and articulating these relations in form of explanations. Whether an individual scientist is able to understand a phenomenon, to grasp relations and articulate explanations, depends on the available body of knowledge, trained research skills, and further material equipment, all of which have to be successfully coordinated and applied. These resources are provided by the scientific community. Once a scientists gained understanding, it needs to be judged whether this understanding is legitimate according to the employed disciplinary norms, whether it is scientific at all. Making this judgment requires a scientific community, again.

Understanding a phenomenon requires having knowledge relevant to the phenomenon, having the research skills and equipment to use this knowledge, to apply it to the phenomenon in the relevant aspects according to one's epistemic goals, in order to articulate an adequate explanation. The explanation is the articulation of the grasped relations, but the understanding itself is a cognitive ability manifested in these processes which cannot be entirely articulated propositionally. In contrast, one can simply know an explanation of a phenomenon from a textbook or testimony without grasping anything about the phenomenon and without constructing an explanation. The crucial difference between explanation and understanding, and also between any form of propositional knowledge and understanding, is that scientific understanding requires, in addition to having knowledge or an explanation, the ability to grasp relations and to use various research skills to make these relations comprehensible and articulate them in an explanation. In the context of scientific research, knowledge of an explanation of a phenomenon is not a first stage that is prior to and separated from the stage of understanding that phenomenon. Rather, explanation is an integral part of understanding, it is constitutive for understanding. Scientists understand phenomena by, with, and through the explanations they construct by employing scientific practices.

The GE-account of scientific understanding addresses various issues with which other accounts of understanding, like the ones provided by Henk de Regt, Kareem Khalifa, and Finnur Dellsén are also engaged. What makes the GE-account distinct from these other accounts, and what are its advantages in comparison to them? I elaborate on these questions in the next section.

## 6.2 Benefits of the GE-account of scientific understanding

What does the GE-account have to say about scientific understanding that has not been sufficiently covered or addressed by other accounts? Does the GE-account provide a more suitable analysis of scientific understanding than other accounts? I argue that it does. In this section, I compare the GE-account of scientific understanding to the accounts developed by Henk de Regt, Kareem Khalifa and Finnur Dellsén. I highlight the weaknesses and problems of these accounts and show how the GE-account is not affected by the issues that the other accounts are facing.

### 6.2.1 Theories are not always crucial for scientific understanding

I start again with Henk de Regt's account of scientific understanding. While a detailed presentation of de Regt's account is provided in section 2.1, let me summarize its most important features. De Regt differentiates between two kinds of understanding which are crucial in science. The first one is UP (understanding a phenomenon) that he characterizes as having an adequate explanation of the phenomenon. An explanation relates the phenomenon to accepted items of knowledge.

De Regt presents this criterion for understanding a phenomenon, which he calls CUP:

> A phenomenon P is understood scientifically if and only if there is an explanation of P that is based on an intelligible theory T and conforms to the basic epistemic values of empirical adequacy and internal consistency.[22]

The second kind of understanding is the understanding of a theory (UT), which means that scientists are able to use the theory. The understanding of a theory is spelled out in terms of intelligibility.

> [De Regt] define[s] the intelligibility of a theory (for particular scientists) as [...] the value that scientists attribute to the cluster of qualities of a theory (in one or more of its representations) that facilitate the use of the theory. It is important to note that intelligibility, thus defined, is not an intrinsic property of a theory but an extrinsic, relational property because it depends not only on the qualities of the theory but also on the skills of the scientists who work with it. Theories are not intrinsically intelligible or unintelligible, but intelligible or unintelligible to a particular scientist or group of scientists. In other words, intelligibility is a context-dependent value.[23]

The thesis that scientists need intelligible theories if they want to gain scientific understanding of phenomena is the basis of de Regt's theory of scientific understanding. If a theory is not intelligible to scientists, they will not be able to use the theory to construct an explanation of a phenomenon on the basis of that theory. Without understanding a theory, understanding a phenomenon is impossible. This implies that de Regt has to determine under which conditions a theory is intelligible. If a theory is intelligible, i.e. if scientists understand the theory, they will have to have some idea of how the theory functions or how it produces certain outputs. Since de Regt allows for a wide variety of theories to provide understanding, he allows for a variety for criteria to assess the intelligibility of a theory. He offers one possible criterion for the intelligibility of theories (CIT):

> **$CIT_1$:** A scientific theory *T* (in one or more of its representations) is intelligible for scientists (in context *C*) if they can recognize qualitatively characteristic consequences of *T* without performing exact calculations.[24]

22 De Regt (2017), p. 92.

23 Ibid. p. 40.

24 Ibid. p. 102.

By including the individual scientists and the specific context, $CIT_1$ accommodates the pragmatic and context-dependent nature of the intelligibility of theories, and, hence, also of UT and UP, since both notions depend on the intelligibility of theories. Furthermore, de Regt argues that understanding cannot be achieved by performing a rule-following procedure. Instead, tacit skills, the know-how to make use of a theory or an explanation, are required. Which skills a scientist needs to make a theory intelligible to her depends partially on the qualities of the theory. By applying $CIT_1$, it is possible to check whether the scientists have developed the appropriate skills for a specific theory. Besides the particular qualities of the theory in question, the combination of established scientific practices in a certain field, the developed abilities or skills of the individual scientists, and the established and available background knowledge determine whether a theory is intelligible for an individual scientist or group of scientists, or not.[25]

The context-dependency of scientific understanding is also crucial for the role of explanation for achieving understanding. De Regt applies a generic conception of explanation, namely that "all explanations are […] arguments […] presenting a systematic line of reasoning that connects [the phenomenon] with other accepted items of knowledge (e.g. theories, background knowledge)."[26] Again, according to de Regt, the construction of explanations on the basis of theories is a matter of skill, of pragmatic decisions which lead to the desired result. He takes understanding to be an epistemic skill. Scientists have to have the know-how to address and solve a new problem. There are no fixed general rules that guide every possible construction process. Various models of scientific explanation, like causal or unificationist explanations, provide different tools for understanding, and all of them may be legitimately used in certain circumstances or contexts. The theory of scientific understanding developed by de Regt accommodates solely explanatory understanding, the understanding that is produced by a scientific explanation.

I agree with de Regt's account in many respects. As I argue throughout this book, I also take understanding to be an ability that includes the articulation, de Regt would say construction, of an explanation of the phenomenon that scientists try to understand. However, I disagree with de Regt in the sense that I do not give theories the central function for scientific understanding that he attributes to them. De Regt uses Ronald Giere's view of scientific theories, according to which scientific theories are "(collections of) principles which provide the basis for the construction of more specific models of parts (or aspects) of the real world."[27] In de Regt's

25 See ibid. p. 103.

26 Ibid. pp. 24f.

27 Ibid. p. 32. For more details concerning Giere's view, see Giere, R. N. (1999), *Science without Laws*. Chicago, University of Chicago Press; and Giere, R. N. (2004), "How models are used to represent reality." *Philosophy of Science*, 71, pp. 742–752, DOI: 10.1086/425063.

view, scientific understanding of phenomena requires intelligible theories. However, he is aware that this centrality of scientific theories for his account of scientific understanding might be problematic and discusses three possible objections. The first one comes from the "new experimentalists", a movement within philosophy of science that started in the 1980s. Key figures of this movement are Ian Hacking, Nancy Cartwright, Deborah Mayo, and also Ronald Giere, among others. New experimentalists claim that a theory-centered perspective on science should be rejected, and that experimentation, instrumentation, and laboratory practices should be analyzed instead, since these activities can be theory-independent. "If this is correct, it would suggest that scientists can achieve understanding without theories: Who would want to deny that scientific experiments provide us with understanding of the phenomenon under investigation?"[28] As a second argument, de Regt considers the claim that philosophical theories of science should not be focused on theories, because theories are comparatively unimportant or not present at all in some scientific disciplines. De Regt considers more descriptive branches of biology, geology, and the social sciences as candidates for scientific disciplines in which theories do not play an important role. "The thesis that theories are essential for achieving scientific understanding seems to entail that these fields and disciplines cannot deliver understanding at all, which obviously would be an unacceptable conclusion."[29] As a third and final argument against the central function of theories for scientific understanding, de Regt considers the claim that the construction of scientific models can be entirely independent from theories, given that models are taken to be autonomous agents. According to this view, scientists would need models, but not theories, to understand phenomena.[30]

De Regt argues that none of the three arguments just presented can accommodate scientific understanding, because "theory is far more pervasive than the objection[s] suggest. Of course, science can be practiced in the absence of full-fledged, explicitly articulated theories, and there is no a priori reason to assume that this cannot lead to (explanatory) understanding."[31] However, if one accepts Giere's liberal conception of theories, which are taken to be (collections of) principles that provide the basis for model construction and experimentation, it is difficult or even impossible to think of science as being theory-independent, so de Regt argues. Scientific experimentation and model-building always take place within a theoretical context and require theoretical interpretation. "Thus, while explicitly articulated theories may be less common in certain areas of geology, biology, psychology, and sociology, scientific activity will still be guided by more loosely circumscribed theoretical princi-

28 De Regt (2017), p. 95.

29 Ibid. p. 95.

30 See ibid. pp. 95f.

31 Ibid. p. 97.

ples."[32] In de Regt's view, theories, low-level or high-level, implicit or explicit ones, are ubiquitous across all scientific disciplines.[33]

It is comprehensible why de Regt insists on the central role of theories for scientific understanding, given the case studies from physics on which he grounds his account of scientific understanding. In these case studies, de Regt investigates the intelligibility of Newton's theory of gravitation from the seventeenth to the nineteenth century, the role of mechanical models in nineteenth century physics, including the effort of understanding the so-called specific heat anomaly on the basis of the kinetic theory of gases and through the dumbbell model provided by Boltzmann, which I discuss in more detail in section 4.2.2, and finally the debates about the intelligibility of matrix mechanics and wave mechanics at the transition from classical physics to quantum physics in the early twentieth century. In all these cases, explicitly articulated theories played a central role for the understanding of physical phenomena. De Regt's notions of UP and UT capture well how physicists achieved understanding in these cases and his analysis provides important insights about the changes of criteria for adequate explanations, intelligible theories, and, therefore, for scientific understanding itself in the course of history. However, de Regt himself states that not only historical, but also disciplinary variation influence the achievement of understanding. And if we look at different disciplines, as I do in chapter five with the episode from biology, it becomes apparent that theories, although not completely absent if Giere's broad notion of theories is adopted, are not always a crucial factor for gaining scientific understanding.

If theories do not play a central role in achieving scientific understanding, they should not be a central concept in any account of scientific understanding. I am not saying that de Regt is completely wrong with his claim that scientists necessarily need theories to understand phenomena. I am saying that he overstates the function of theories for scientific understanding, which is due to his focus on only one discipline that may be viewed as the prime example for a scientific discipline employing explicitly articulated theories: physics. In the episode on the research on zebrafish, Gritsman and her colleagues do not even mention the term 'theory' in the paper in which they present their understanding of the function of the Oep protein. This does not mean that the genetic theory was completely absent in this episode or that the genetic theory was not intelligible for the scientists working with zebrafish. Yet, the importance of the genetic theory to understand the function of the Oep protein was insignificant in comparison to the knowledge of experimental data and results that the scientists obtained. Whatever your favorite philosophical account of a scientific theory may be, the biologists researching the Oep protein did not attribute any crucial function to the genetic theory or other theories that may implicitly have

32 Ibid. p. 97.

33 See ibid. pp. 97ff.

shaped the understanding. Therefore, I do not include any explicit notion of theory in the GE-account of understanding. Instead, I subsume theories under the notion of propositional knowledge, together with laws, principles, axioms, or data. Each of these propositional concepts can play a role for scientific understanding.

The GE-account of scientific understanding can accommodate scientific disciplines in which explicitly articulated theories are less common or not as crucial for scientific understanding as other information better than de Regt's account of scientific understanding. At the same time, the GE-account can also capture those disciplines that employ explicitly articulated theories, like physics. To understand a phenomenon, scientists need knowledge, and this knowledge may comprise knowledge of theories, laws, principles, or empirical data. Which knowledge the scientists need to possess precisely depends on the phenomenon they are trying to understand and on the disciplinary and historical context. By not putting a heavy emphasis on theories, the GE-account has a greater flexibility in accommodating various scientific disciplines. As de Regt highlights himself, "as long as the general characterization and criteria for understanding include elements that allow for historical and disciplinary variation, it is perfectly well possible to formulate an account that transcends the purely local context."[34] By not giving theories a center stage in an account of scientific understanding, the GE-account allows for an additional dimension of variation across historical and disciplinary contexts that de Regt's account cannot offer.

What is more, the GE-account avoids possible criticism that can be raised against de Regt's and Giere's notion of theories as collections of principles. The philosophical debate on theory itself is a huge one, and several conceptions of theories exist. There is not only disagreement between proponents of the Syntactic View (defining theories as axiomatized collections of sentences), the Semantic View (taking theories to be collections of nonlinguistic models) and the Pragmatic View (according to which theories are amorphous entities possibly consisting of sentences and models, and additionally of problems, skills and practices), but also among proponents of one and the same view concerning its details.[35] There is no consensus about how scientific theories should or could be conceptualized. One can argue against Giere, whose conception of a scientific theory falls under the Semantic View, and de Regt that theories are something else than (collections of) principles. The GE-account does not face this issue at all. If scientists in a certain episode understand a phenomenon through or with the help of a theory, which the scientists themselves view as a theory, then this theory will play an important role

34 Ibid. p. 11.

35 For more information about the different views on scientific theories, see Winther, R. G., "The Structure of Scientific Theories", *The Stanford Encyclopedia of Philosophy* (Spring 2021 Edition), Edward N. Zalta (ed.), URL = https://plato.stanford.edu/archives/spr2021/entries/structure-scientific-theories/ (last accessed April 14th, 2022).

in the understanding. However, if scientists in another episode do not refer to or mention a theory at all, then theories will at least not play the most important role for understanding, even if some theory might have influenced the understanding in an indirect way. Again, since the Practice Turn in the philosophy of science, it has been criticized that philosophers of science have put too much emphasis on the concept of theory in the past. Instead of insisting on a central function of theories for scientific understanding and identifying a theory in any episode at any cost, scientists' own views about theory, explanation, and understanding should be taken into account. Whether understanding is approved as scientific should not depend on any theory on which the understanding may be based, but rather on the method by which it is achieved. A method that is governed by the rules and standards implemented in the respective historical and disciplinary context.[36]

I would like to point to a second issue of de Regt's account of scientific understanding. As already mentioned, he distinguishes between UT (understanding a theory) and UP (understanding a phenomenon). Although de Regt claims that he is concerned with UP, he says comparatively little about it. His account rather focusses on UT, "the (pragmatic) understanding of the theory that is used in the explanation"[37], the procedures through which physicists in various historical episodes use theories to construct explanations. In contrast to this demanding and challenging process, UP, characterized as having an adequate explanation of the phenomenon, does not seem to be very impressive. What's more, this characterization of UP causes irritation. De Regt emphasizes again and again that knowing an explanation is not sufficient for or identical to understanding. The question then arises what he means when he says that a scientist *has* UP if she *has* an adequate explanation of the phenomenon. Does 'having an adequate explanation of the phenomenon' not mean that one possesses, or knows, an adequate explanation of the phenomenon? If this is the case, de Regt would contradict his own claims. Granted, this contradiction would only affect UP, not UT. De Regt could defend his account by arguing that UT is an epistemic skill, which he shows convincingly. Still, this would imply that UP is not an epistemic skill. One kind of understanding, UT, is a skill, but another kind, UP, is not? If UT is conceptualized as a skill and UP is not, they are completely different things. Why are both concepts then labelled understanding?

I do not think this interpretation captures de Regt's opinion on understanding. I think he takes UT and UP to be epistemic skills, the former serving the latter. For de Regt, understanding a theory means that scientists have the skills to use the theory to construct explanations. UT is the means to achieve UP, which he takes to be the aim and product of scientific explanations. A scientific explanation provides a

36 How theories might contribute to scientific understanding is discussed, among others, in de Regt (2017) or Baumberger & Brun (2017).

37 De Regt (2017), p. 24.

systematic line of reasoning through which a phenomenon is connected to accepted empirical and theoretical knowledge. When this connection is established, scientists can apply, refine and extend their knowledge further.[38]

> This is the sense in which the explanation provides understanding of the phenomenon. This can be illustrated [...] with the example of the kinetic theory of gases. Elementary phenomenological gas laws can be explained on the basis of the kinetic theory by constructing the ideal-gas model [...]. The intelligibility of the kinetic theory is a precondition for constructing the model-based explanation of the phenomenological law. But in what sense does this explanation provide understanding? The answer is: by connecting our empirical knowledge of gaseous behavior with accepted theoretical knowledge (in this case, e.g., with Newtonian mechanics) the explanation allows us to make inferences about the behavior of gases in novel situations, and to extend, apply, and refine our knowledge. [...] The crucial point is that the skills that are required for constructing and evaluating an explanation are the same as those required for using and extending it. And, as I have argued, it is precisely the possible use and extension of an explanation that embodies the understanding that comes with it. In other words, understanding of the theories on which the explanation is based (UT) corresponds in a fundamental way with the understanding generated by the explanations (UP).[39]

Although de Regt's idea that UP is embodied in the possible use and extension of an explanation is very similar to, and probably compatible with, my idea of articulating new explanations, he does not elaborate this idea very clearly. In the three case studies that de Regt presents, he is primarily concerned with UT, the ability of scientists to use theories. He examines first the intelligibility of Newton's theory of universal gravitation in the seventeenth century, second the use of mechanical models in nineteenth century physics to explain, for example, the specific heat anomaly through the kinetic theory of gases, and third the debate about the intelligibility of matrix mechanics and wave mechanics in early twentieth century quantum mechanics. Since de Regt's account of scientific understanding essentially relies on the thesis that explanatory understanding of phenomena requires intelligible theories, he analyzes in detail the context-sensitive, de Regt says context-dependent, nature of the intelligibility of theories, and which skills and conceptual tools scientists need so that a theory is intelligible to them in different scientific episodes. I do not disagree on any of the points that de Regt presents, but I want to highlight that his analysis is extremely focused on the construction and evaluation of explanation (UT), and not so much on using and extending explanation (UP). In the case studies, de Regt argues at length which theory was intelligible or not to which scientists and who was able to use a

38 See ibid. pp. 44ff.

39 Ibid. p. 46.

theory to construct an explanation of a phenomenon based on the theory in question. His analysis sometimes reads as if UT is a process or procedural ability, while UP is a result, a state, that is reached or not. Other times, de Regt writes that UP is "relating the phenomenon to accepted items of knowledge"[40], but does this framing not overlap with UT as the ability to construct explanations? In short, while de Regt's analyses of how physicists in different historical episodes gained explanatory understanding of phenomena on the basis of intelligible theories are very illuminating, it is not ultimately clear what UT and UP consist in, respectively, and how exactly they relate to each other. The GE-account offers an alternative, and possibly complementary, perspective on scientific understanding of phenomena. By taking the understanding of the available body of knowledge (including theories) for granted, the GE-account concentrates on the understanding of phenomena. This avoids possible irritation as to when UT or UP are the topic of analysis. By focusing on one single concept of scientific understanding defined as one ability, manifesting in the iterative process of grasping relations of the phenomenon and articulating them in form of explanations, it provides a starting point to rethink the notions of UT and UP, and their relation.

In sum, de Regt's account and the GE-account of scientific understanding agree in many respects. However, de Regt's account might face problems when it is used to accommodate episodes from scientific practice in which theories are either not present at all or do not play a central role in the manifestation of understanding in comparison to other pieces of knowledge or information. Additionally, the acceptance of de Regt's account of scientific understanding stands and falls with the conception of scientific theories one accepts. The GE- account avoids these problems completely, since I neither adopt a specific conception of scientific theories, nor do I argue that understanding of phenomena always requires theories. Furthermore, since de Regt differentiates between UT and UP and elaborates a lot on UT, but not so much on UP, it does not become clear in his analysis what exactly UP is, what it consists in, and what it adds to UT. This is a central and pressing issue, since UP is the ultimate aim that scientists want to achieve, as de Regt states himself. As the GE-account only covers understanding of phenomena, it can be seen as complementary to de Regt's theory.

### 6.2.2 Neither grasping nor understanding simply amount to knowledge

Kareem Khalifa has developed a different model of scientific understanding. As in the previous section, I first provide a summary of Khalifa's account of understand-

40 Ibid. p. 91.

ing, before I address its deficiencies.[41] A more detailed presentation of Khalifa's account can be found in section 2.2.

Khalifa calls his account EKS model of understanding (explanation, knowledge, science model), because these three concepts are crucial for his account of better understanding:

> ($EKS_1$) $S_1$ understands why *p* better than $S_2$ if and only if:
> (A) *Ceteris paribus*, $S_1$ grasps *p*'s explanatory nexus more completely than $S_2$; or
> (B) *Ceteris paribus*, $S_1$'s grasp of *p*'s explanatory nexus bears greater resemblance to scientific knowledge than $S_2$'s.[42]

Additionally, the model includes a third principle, $EKS_2$, which accounts for minimal understanding. $EKS_2$ answers the question under which conditions someone has any understanding of a phenomenon, which is not equal to understanding achieved through scientific research.

> ($EKS_2$) *S* has minimal understanding of why *p* if and only if, for some *q*, *S* believes that *q explains why p*, and *q explains why p* is approximately true.[43]

Significantly, Khalifa follows the "received view" of understanding, as he calls it, which states that understanding is a kind of knowledge of explanation. "*S* understands why *p* if and only if there exists some *q* such that *S* knows that *q explains why p*."[44] Therefore, his model, as de Regt's account, is only concerned with explanatory understanding, the understanding-why something is the case.

The first principle labelled (A) in $EKS_1$ is called Nexus Principle. Khalifa starts with the idea that the subject's understanding of a phenomenon increases if she knows more correct explanatory factors that contribute to the phenomenon and if she knows more of the relations that exist between these factors. On this basis, Khalifa defines the explanatory nexus of a phenomenon *p* as "the set of correct explanations of *p* as well as the relations between those explanations."[45] If the explanatory nexus of *p* only includes correct explanations, then what counts as a correct explanation? Khalifa presents these four conditions:

---

41 For a further critique of Khalifa's account, see De Regt, Henk W. and Höhl, Anna E. (2020), Review of Khalifa, K., *Understanding, Explanation and Scientific Knowledge*, BJPS Review of Books, https://www.thebsps.org/reviewofbooks/kareem-khalifa-understanding-explanation-and-scientific-knowledge-reviewed-by-de-regt-hohl/ (last accessed April 14th, 2022).

42 Khalifa (2017b), p. 14.

43 Ibid. p. 14.

44 Ibid. p. 18.

45 Ibid. p. 6.

*q* (correctly) explains why *p* if and only if:
*(1)* *p* is (approximately) true
*(2)* *q* makes a difference to *p*
*(3)* *q* satisfies your ontological commitments (so long as they are reasonable); and
*(4)* *q* satisfies the appropriate local constraints.[46]

The fourth condition is important. Like de Regt, Khalifa explicitly allows for an explanatory pluralism. He does not give a strict definition of explanation. In fact, he even allows to identify 'explanation' with 'explanatory information'.[47] With local constraints he refers to the specific interest of the researcher, the established standards of the discipline, and so on. Local constraints are context-dependent. Like de Regt, Khalifa wants to formulate an account of understanding that is universally valid, but allows for contextual variation. Khalifa reaches this goal by formulating three global conditions and one local condition for explanation.

The second principle contained in $EKS_1$ is the Scientific Knowledge Principle. This principle captures everything Khalifa takes to be necessary for a characterization of grasping. He defines grasping as "a cognitive state bearing some resemblance to scientific knowledge of some part of the explanatory nexus."[48] But what counts as scientific knowledge? Knowledge is scientific if it has been gained through *scientific explanatory evaluation*, SEEing, in short. According to Khalifa, SEEing consists of three components: the consideration of plausible potential explanations of the phenomenon of interest, the comparison of the potential explanations, and finally of the formation of (doxastic) attitudes based on the comparisons. SEEing ensures the safety of one's explanatory commitments and therefore the status of this kind of knowledge as scientific.[49]

I agree with Khalifa's model of understanding in so far as I also think that some knowledge about a certain phenomenon is necessary to understand that phenomenon (how could you ever understand any phenomenon without knowing anything about it, not even that it exists?), that explanations are necessary for scientific understanding and that an explanatory pluralism should be adopted in order to accommodate historical and disciplinary variations in science. Khalifa, de Regt, and I have a common ground in this regard. However, I disagree with Khalifa in his claim that understanding is knowledge of an explanation. The fundamental problem I see in Khalifa's account of understanding is his deflationist conception of grasping. In his view, "talk of grasping can always be replaced by a more specific

46 Ibid. p. 7.
47 See ibid. p. 6.
48 Ibid. p. 11.
49 See ibid. pp. 12f.

epistemic status (e.g., approximately true beliefs, non-scientific knowledge, scientific knowledge). In other words we can always swap out the placeholder – the buzzword "grasping" – with something more pedestrian and informative."[50] The term 'grasping', according to Khalifa, has no meaning. It does not denote anything in the world. But, why should grasping merely be a placeholder or a buzzword for other epistemic statuses? Khalifa cannot answer this question, because he does not give any argument or justification for his deflationist view. Admittedly, I do not have an ultimate proof that Khalifa's deflationist view is wrong. However, it should be noted that, to my knowledge, Khalifa is the only one in the philosophical debate on understanding who holds such a view with respect to grasping. As I explain in section 4.3.1, there is no universal agreement in the debate on understanding about what grasping amounts to. The two main options on the market are either to view grasping as a process of getting epistemic access, to recognize or becoming aware of (aspects of) a phenomenon, which is the view that I endorse as well, or to spell out grasping in terms of other reasoning or inferential abilities. Both views are far more demanding and definitely do not take grasping to be only a placeholder for believe or knowledge states. Having these positions concerning grasping in mind, the burden of proof lies on Khalifa's side. As long as neither he nor anyone else provides convincing arguments or evidence for the deflationist conception of grasping as being merely a placeholder, there is little or no reason to accept it.

A second issue that I have with Khalifa's model of understanding relates to his deflationist conception of grasping. As he takes grasping to be only a placeholder term that denotes other states of believe or knowledge and that does not relate to any capacity or ability, he also denies that understanding requires any "special abilities" that are not required for explanatory knowledge. However, Khalifa does acknowledge the role and importance of skills for achieving scientific knowledge through SEEing.

> Note that each aspect of SEEing involves significant cognitive abilities. For instance, consideration involves highly structured creativity (when generating alternative explanations). Comparison involves insight into different explanatory relationships (e.g. causal structures, dependency relationships, inferential connections within and between explanations), the ability to draw out predictive consequences of each explanation, and various kinds of methodological prowess, such as the ability to design experiments and interpret results. Formation deploys inferential abilities.[51]

50 Ibid. p. 14.

51 Ibid. p. 63. See also ibid. chapter three for more details of Khalifa's view that understanding does not require special abilities.

Khalifa views the product of the process of SEEing to be scientific explanatory knowledge, alias understanding, whereas I take the process of SEEing, the process and activity of creating scientific explanatory knowledge, to be the manifestation of the disposition to understand a phenomenon. The disagreement between Khalifa and me already starts with his definition of minimal understanding.

> ($EKS_2$) *S* has minimal understanding of why *p* if and only if, for some *q*, *S* believes that *q explains why p*, and *q explains why p* is approximately true.

I disagree that (minimal) understanding is a form of belief. I may believe that the global mean surface temperature on Earth increases because of a higher concentration of greenhouse gases in the atmosphere, but having this belief does not enable me to understand in any sense how or why the rising concentration of greenhouse gases in the atmosphere lead to an increase the global mean surface temperature. Believing this explanation does not entail any abilities to recognize how the global mean surface temperature and the greenhouse gas concentration in the atmosphere are related or how changes in the earth-atmosphere-system may influence the phenomenon of rising mean surface temperatures. Basically, a person may have a lot of knowledge or many beliefs about various aspects of global climate change, but she may never be able to grasp, to recognize or to comprehend, how these various pieces of knowledge relate to each other and how they relate to actual phenomena in the world. A belief or knowledge about a phenomenon is a necessary prerequisite for understanding the phenomenon, and some explanatory knowledge is the product of understanding, but understanding itself is not identical to a belief or knowledge.

As I argue in chapter four, the concepts of propositional knowledge (knowing-that) and understanding (knowing-how) can be easily confused because they both advance only in conjunction with one another. Having understanding denotes the ability to make sense of a certain phenomenon or a specific observation by referring to, using, manipulating and coordinating the newly gained insights or information concerning the phenomenon or observation with already available (background) knowledge through various possible cognitive or material skills. A scientist understands a phenomenon if she is able to align new insights about a phenomenon (new observations or new data obtained in an experiment or study) with the available background knowledge. In the course of this process, the scientist who understands the phenomenon will articulate and provide a new explanation of the respective phenomenon that will be integrated in the existing body of knowledge if parts of the wider scientific community accepts the new explanations as valid. New knowledge of an explanation of a phenomenon is a result of understanding this phenomenon, but it is not identical to understanding. This complex ability exceeds any notion of belief or knowledge by far.

Furthermore, notice that Khalifa may face a different problem here. He argues that understanding is knowledge of an explanation. However, according to his definition of minimal understanding, understanding is having a belief about an explanation. A belief is not identical to knowledge, it is even less in terms of epistemic demands. As I see it, Khalifa has two options. Either he has to say that understanding is believing an explanation, and the understanding improves in terms of better justification of that belief or by approximating truth in some way, or he has to modify his definition of minimal understanding as having a minimally justified belief about an explanation. Otherwise Khalifa identifies knowledge with belief, and I cannot imagine him seriously advocating this claim.

Summing up, Khalifa and I also agree in many respects. Both of us acknowledge the crucial role of demanding abilities in the process of scientific research of a phenomenon and the articulation of scientific explanation in the course of this process. However, we fundamentally disagree in our conceptions of understanding. Whereas *Khalifa views understanding as the product of a research process*, the articulated scientific knowledge of an explanation, *I take the whole process* of grasping relations and articulating an explanation of the phenomenon, for which scientists have to generate and test hypotheses, construct models, using various research methods and evidence, *to be the manifestation of understanding*. My argumentation in this book that understanding is an ability and not a kind of knowledge may not convince Khalifa or any other proponent of the 'understanding is a kind of knowledge'-camp due to incompatible intuitions regarding understanding. Nevertheless, I am convinced that an ability-account of understanding is better suited to do justice to the demanding epistemic activity of gaining understanding, in a scientific as well as non-scientific context.

### 6.2.3 Why grasping is not enough for understanding

Lastly, I would like to compare the GE-account with the account from Finnur Dellsén. His account differs significantly from the accounts from de Regt and Khalifa, since Dellsén argues for understanding without explanation, for an account of objectual understanding. Again, let me quickly repeat the most important characteristics of Dellsén's account of objectual understanding before I compare it to the GE-account of understanding.

As the previously mentioned two scholars, Dellsén is interested in the understanding of phenomena and assumes that typical cases of this sort of understanding can be found in the sciences. Hence, his account is intended to capture scientific understanding. Moreover, Dellsén is also convinced that understanding is gradual in a way knowledge is not. In his view, scientists have to grasp a model of a phenomenon's dependence relations if they want to understand the phenomenon. Dellsén takes models to consist of two components, namely some kind of information structure and an interpretation, which relates elements of the information structure to elements of the phenomenon. In a nutshell, "understanding consists of grasping a certain kind of model of the understood phenomenon"[52], according to Dellsén. More precisely, scientists must grasp a model that represents the dependence relations that the phenomenon stands in towards other things. That is, scientists must grasp a dependency model. As models are always incomplete representations of their targets, as they are not copies, the quality of a dependency model can vary along two different dimensions, according to Dellsén, which are accuracy (tied to idealization or the misrepresentation of some features) and comprehensiveness (tied to abstraction or the omission of some features). Since both criteria, accuracy and comprehensiveness, are gradable notions, the degree of understanding will depend on the degrees of the accuracy as well as the comprehensiveness and the trade-off between the two regarding any dependency model that is grasped.[53]

In short, Dellsén proposes the following dependency modelling account (DMA) of understanding:

> DMA: *S* understands a phenomenon, *P*, if and only if *S* grasps a sufficiently accurate and comprehensive dependency model of *P* (or its contextually relevant parts); *S*'s degree of understanding of *P* is proportional to the accuracy and comprehensiveness of that dependency model of *P* (or its contextually relevant parts).[54]

DMA does not require explanation, although Dellsén takes dependence relations to usually undergird explanations. He contrasts his DMA with explanatory accounts of understanding, which he summarizes in the following way:

> *U→E*: *S* understands *P* only if *S* grasps enough of an adequate explanation of *P* (or its relevant features); other things being equal, *S* has more understanding of *P* to the extent that *S* grasps more of an adequate explanation of *P* (or its relevant features).[55]

*U→E* is intended to capture any account of explanatory understanding that takes explanation as a necessary requirement for understanding. Dellsén then discusses three cases in which, according to him, *U→E* fails to accommodate the understanding that scientists achieve, whereas DMA can cope with such types of cases. Before I turn to these cases, I would like to address *U→E* and its relation to the GE-account of understanding.

---

52 Dellsén (2020), p. 1265.

53 See ibid. pp. 1266ff.

54 Ibid. p. 1268.

55 Ibid. p. 1269.

I do claim in the GE-account that explanation is a necessary requirement for understanding a phenomenon. However, I do not claim that *grasping an explanation* of a phenomenon is a necessary requirement for understanding it. Scientists grasp relations of the phenomenon and articulate what they have grasped in form of an explanation. Therefore, the GE-account is, strictly speaking, not included in *U→E* and may not be affected by Dellsén's criticism of explanatory accounts of understanding. Unfortunately, Dellsén does not clarify what exactly he takes explanations to be, whether he holds an ontic or epistemic conception of explanation. Yet, since he differentiates between explanations and dependence relations that undergird explanation, it seems more plausible to attribute an epistemic conception of explanation to Dellsén. If this is correct, it becomes questionable, though, whether *U→E* does capture most or all accounts of explanatory understanding, as Dellsén wants it to be. Consider two explanatory accounts that he explicitly mentions and takes to be comprised by *U→E*, the accounts of Michael Strevens and Henk de Regt.[56] Both argue for explanatory accounts, but hold completely different conceptions of explanation. Strevens advocates an ontic conception of explanation, de Regt an epistemic conception. Granted, since de Regt characterizes understanding of phenomena as having an adequate explanation, it is comprehensible why Dellsén takes his account to be covered by *U→E* as well. Yet, as I argue in section 6.2.1, this formulation is very unfortunate and does not really capture what de Regt takes scientific understanding to be. Independently of any interpretation of de Regt's account of understanding, the point I want to make here is whether *U→E* succeeds in capturing all explanatory accounts of understanding. Taken for granted that some accounts employ an ontic conception of explanation and others an epistemic conception, what exactly is it that subjects grasp according to these accounts? Do they grasp explanations, because, according to the ontic conception, explanations are out there in the world, or do they grasp (dependence) relations, and then, as maintained by the epistemic conception, articulate or construct explanations of these relations? These are two very different activities, as long as the conception of 'grasping' is not broadened in a way that it also captures the activities of articulating or constructing explanations. Christoph Baumberger, for example, presents such a wider conceptualization of grasping that I address in section 4.3.1.

The upshot of the discussion of *U→E* in the previous paragraph is that it should be made clear what is meant by the term 'explanation' and that, depending on that meaning, any definition or characterization of (explanatory) understanding may fundamentally change. Therefore, it is questionable whether *U→E* does capture most

56 De Regt's account of understanding is discussed at length in sections 2.1 and 6.2.1. For more information on Strevens' account, with which I do not engage in more detail, see Strevens (2013).

or all explanatory accounts of understanding, including the GE-account of understanding. However, as Dellsén compares *U→E* with DMA in terms of three cases, turning to them may shed more light on how *U→E* and DMA might differ in Dellsén's view. On the next pages, I argue that the GE-account can better accommodate the examples that Dellsén presents than his very own DMA.

#### 6.2.3.1 Understanding the values of dimensionless physical constants

The first type of cases concerns 'explanatory bruteness', as Dellsén calls it. 'Explanatorily brute' facts "are phenomena that have no explanation at all – phenomena that are not merely unexplained, but unexplainable."[57] Everyday coincidences or fundamental physical truths, like the values of dimensionless physical constants, are instances of explanatorily brute facts. The fine structure constant or Sommerfeld's constant $\alpha=1/137$, which describes the strength of electromagnetic interaction between elementary charged particles, is a dimensionless physical constant. These constants cannot be explained by any current physical theory, they can only be measured. Assuming that there are indeed no explanations for the values of dimensionless physical constants, Dellsén argues that DMA can easily accommodate such cases of explanatorily brute facts, "since a dependency model that depicts such a phenomenon or its features as not dependent on anything else would be more accurate than an otherwise identical model that represents them as dependent on something else, and more comprehensive than an otherwise identical model that abstracts away from the issue."[58] Accordingly, a scientific community that discovers and accepts that these values are explanatorily brute is better off than a scientific community that is still wondering whether the values can be explained, according to Dellsén.[59]

Dellsén's claim that DMA is superior to explanatory accounts of understanding, since it can accommodate cases of explanatory bruteness, is not convincing, because, as Dellsén himself admits, "it is very much an unsettled empirical question whether a given fact is explanatorily brute."[60] That is, there is no *a priori* reason to assume that facts which we cannot explain, yet, like the values of dimensionless physical constants, are explanatorily brute. Just because we cannot explain a

---

57 Ibid. p. 1271.

58 Ibid. p. 1275.

59 See ibid. pp. 1274f. Dellsén also discusses at length Kvanvig's example of the moving electron, as a special type of explanatorily brute facts, and Khalifa's criticism of Kvanvig's interpretation of the case, see ibid. pp. 1271–1274. I do not have the space to go into Dellsén's discussion of this specific case. However, since I argue against his general claim that understanding of explanatorily brute facts is possible, it is not necessary to go into this specific case, too. For a detailed discussion of Kvanvig's example and Khalifa's response, see section 3.3 in this book.

60 Ibid. p. 1274.

phenomenon, yet, does not automatically amount to the conclusion that this phenomenon is not explainable. I am not denying that explanatorily brute facts may exist in the universe, but I claim that we can never know for sure whether any as yet unexplained fact is explanatorily brute. Taking the incredible amount of scientific discoveries of phenomena into account, it is reasonable to assume that future science will be able to explain phenomena that cannot be explained, yet. And it is not possible to know in advance which phenomena we will be able to explain in the future. Dellsén's claim that a scientific community that takes values of dimensionless physical constants to be explanatorily brute has a better understanding as a scientific community who does not is correct just in case these values are indeed explanatorily brute. If this is not the case, it will be the other way around and the first scientific community will never understand these values.

Furthermore, why should we want an account of understanding that covers instances of explanatory brute facts in the first place? Why should this be an advantage? What is so bad about admitting that we can never understand explanatorily brute facts, while we can understand multiple phenomena that are related to these facts? As Khalifa states, "certain information helps to provide (explanatory) understanding of something else, even if it is not itself understood."[61] It is not necessary, and probably also not possible, to understand every dependency model, body of knowledge, or even every single dependence relation within our grasp. Dellsén admits this as well, as in his view "context plausibly determines which parts of a complex phenomenon need to be understood to a significant degree in order for it to be felicitous to say that the phenomenon itself is understood."[62] And even if we consider a context in which some hitherto unexplained fact, like the value of a dimensionless physical constant, needs to be understood for whatever reason, it is possible to make a normative claim of why physicist should strive to find an explanation of this value. Given that there is no proof of the existence of explanatorily brute facts generally, nor of the explanatory brute nature of any one specific fact, scientists should strive for finding an explanation of the respective fact. Maybe no explanation will be found, maybe it will be proven that this fact is explanatorily brute, but the (re)search for an explanation will very likely lead to new discoveries that cannot be imagined, yet. The strive for understanding, but also explanation, of phenomena is the engine of scientific progress. Accepting an unexplained fact as explanatorily brute without having a reason or explanation for this decision may prevent this progress and undermine the very nature of science. In fact, physicists are trying for decades to find explanations for the values of dimensionless constants. In case of the fine structure constant, Arthur Eddington and Wolfgang Pauli were among those who tried to ex-

61 Khalifa (2013), p. 1166.

62 Dellsén (2020), p. 1268.

plain and understand its value.[63] A famous quote that is very often stated in this context comes from Richard Feynman, who wrote in 1985 "immediately you would like to know where this number for a coupling [the value of the fine structure constant] comes from: is it related to pi or perhaps to the base of natural logarithms? Nobody knows. It's one of the greatest damn mysteries of physics: a *magic number* that comes to us with no understanding by man."[64]

This quick excursion into debates in physics clearly shows that even if DMA captures some kind of understanding of hitherto unexplained, or possible unexplainable facts, this will not be a type scientific understanding with which scientists are satisfied or that they aspire. Quite the contrary, the attempts of and struggle for physicists to find explanations for the values of dimensionless constants, even though they were not successful, yet, demonstrate that a type of understanding characterized by DMA should be overcome and replaced by explanatory understanding. The GE-account of understanding comprises the need for explanation that one recognizes if scientific practice is taken into account. Therefore, the GE-account of scientific understanding is better in accommodating scientific practice than DMA. Consequently, scientists do not (scientifically) understand unexplained or unexplainable facts, and there is absolutely no problem in admitting that scientists do not understand everything. If this were the case, no research would be conducted anymore.

#### 6.2.3.2 What does Bohr's model explain?

A second type of cases in which understanding is achieved without explanation is called 'explanatory targetedness' by Dellsén. "In these cases, we come to understand through grasping an explanation, but the explanation helps us understand the explanans rather than the explanandum. Thus, in these cases, the target of one's understanding differs from the target of one's explanation in a way that separates understanding of *P* from grasping an explanation of *P*."[65] The concrete example dis-

63 For more information concerning research and controversies on the fine structure constant, see for example Whittaker, E. (1945), "Eddington's Theory of the Constants of Nature." *The Mathematical Gazette*, 29 (286), pp. 137–144, DOI: 10.2307/3609461; or Kragh, H. (2003), "Magic Number: A Partial History of the Fine-Structure Constant." *Archive for History of Exact Sciences*, 57 (5), pp. 395–431, DOI: 10.1007/s00407-002-0065-7; or Várlaki, P., Nádai, L., Bokor, J. (2008). "Number archetypes and 'background' control theory concerning the fine structure constant". *Acta Polytechnica Hungarica*, 5 (2), pp. 71–104. For a relatively recent suggestion of an anthropic explanation for the value of the fine structure constant, see Barrow, J. D. (2001), "Cosmology, Life, and the Anthropic Principle". *Annals of the New York Academy of Sciences*, 950 (1), pp. 139–153, DOI: 10.1111/j.1749-6632.2001.tb02133.x.

64 Feynman, R. P. (2006 [1985]), *QED: The Strange Theory of Light and Matter*. Princeton, Princeton University Press, p. 129, original emphasis.

65 Dellsén (2020), p. 1275.

cussed by Dellsén is the transition from Rutherford's planetary model of the atom to Bohr's quasi-quantum model. While both models depict the atom as having a positively charged nucleus that is orbited by negatively charged electrons, the Rutherford model does not designate which locations or energy levels could be occupied by electrons. In contrast, Bohr's model does specify the electron orbits with certain fixed radii that correspond to particular energy levels. Although both models are not accurate representations of the atom and deficient in comparison to the contemporary fully quantum mechanical model, it is agreed that Bohr's model is more accurate than the Rutherford model. Hence, Dellsén takes it as intuitive to say that Bohr's model increased understanding of the atom in comparison to the earlier Rutherford model.[66]

One advantage of Bohr's model in comparison to Rutherford's model is its capacity to provide information through which the Rydberg formula for spectral lines of several elements can be explained. Since electrons can only occupy specific radii (energy levels), when they 'jump' between orbitals they gain or lose energy exclusively in fixed discrete quantities that represent the differences between two radii. This information explains why atoms emit radiation with certain fixed wavelengths described by Rydberg's formula. The Rutherford model could not be used to explain the wavelengths of spectral lines, because it does not entail fixed electron radii and cannot account for the observation of discrete wavelength. This example, according to Dellsén, might suggest that explanatory accounts of understanding can capture the increase of understanding achieved by the transition from Rutherford's to Bohr's model.[67] "But this tempting line of thought is mistaken. To see why, note that the spectral patterns described by Rydberg's formula are not a feature of any atom, but a feature of the radiation that is omitted from such atoms. So the phenomenon that is being explained in the above explanation—the explanandum—is not a feature of the atoms as described by Bohr's model at all."[68]

The explanation Dellsén is concerned with here is the following: the radiation from atoms has certain fixed wavelengths described by the Rydberg formula, because electrons within atoms can only occupy specific energy levels. The information from Bohr's model figures into the explanans, but not in the explanandum. While the model enabled explanation of the atom's spectral pattern, it did not enable understanding of the atom itself, since it merely stipulated features like the fixed electron radii. Therefore, Bohr's model did not increase explanatory understanding of the atom in comparison to Rutherford's model, so Dellsén argues. Unsurprisingly, he claims that DMA can better accommodate this case.[69]

---

66 See ibid. pp. 1275f.

67 See ibid. p. 1276.

68 Ibid. p. 1276.

69 See ibid. pp. 1276f.

> The transition from Rutherford's model to Bohr's provides a more comprehensive model of the dependence relations in which the atom stands towards spectral lines. In this way, DMA validates the judgement that Bohr's model really did increase our understanding of the atom, despite the fact that the model did not provide an explanation of any of the atom's features.[70]

It is true that Bohr wanted to understand the nature and structure of atoms, that his model accurately explains the wavelength of spectral lines emitted by atoms, and that Rutherford's model could not explain this phenomenon. However, Dellsén does not mention another important aspect of this scientific episode. Bohr did not develop his model of the atom because he primarily wanted to explain the wavelength of spectral lines, although this achievement may be viewed as the greatest success of the model, but because Rutherford's model faced other severe problems. Since Rutherford still adhered to the laws of classical mechanics, electrons in his model constantly lose energy in form of electromagnetic radiation (light) while they are orbiting the nucleus. This hypothesis has two problematic consequences. First, as the electrons are constantly losing energy, atoms should emit a continuous stream of electromagnetic radiation as they are spiraling inwards towards the nucleus. Dellsén already described the observation that atoms do not emit a continuous spectrum, but instead light of specific discrete frequencies. Second, and this is presumably the more devastating consequence of Rutherford's model, since electrons are constantly losing energy and spiral towards the nucleus, they will ultimately collapse into the nucleus. This means that no atom is stable! Obviously, this cannot be true, since stable matter exists in various forms. In the publication in which Bohr presents his model for the first time, he wrote in the introduction that in the "attempt to explain some of the properties of matter on the basis of [Rutherford's] atom-model we meet, however, with difficulties of a serious nature arising from the apparent instability of the system of electrons. [...] Whatever the alteration in the laws of motion of the electrons may be, it seems necessary to introduce in the laws in question a quantity foreign to the classical electrodynamics, i.e., Planck's constant".[71] Though Bohr mentions the explanation of the hydrogen spectrum through his model as well at the end of the introduction to this paper, this does not seem to be his motivation or driving question for developing his model. That is, the actual explanatory target of Bohr's model is the stability of atoms, and not the emission of spectral lines described by the Rydberg formula, as Dellsén argues.

70 Ibid. p. 1277.

71 Bohr, N. (1913). "On the Constitution of Atoms and Molecules, Part I". *Philosophical Magazine*, 26 (151), pp. 1–24, pp. 1f. For more details concerning the development of Bohr's model and the historical context, see Robertson, D. S. (1996), "Niels Bohr – Through Hydrogen Towards the Nature of Matter." In Lakhtakia, A. (ed.), *Models and Modelers of Hydrogen*, pp. 49–82, Singapore, World Scientific Publishing.

So, Bohr proposed his model primarily to avoid and solve problems that earlier models of the atom were facing. He had reasons to make the postulations that he did and to introduce a first quantum mechanical interpretation of the atom, since his model presented a stable atom and was in accordance with the early quantum theory of his time. His model can explain spectral lines as well as the stability of atoms. According to Bohr's model, atoms are stable because electrons emit radiation only when they 'jump' between stationary orbits, but not while revolving in one stationary orbit around the nucleus. In his discussion of Rutherford's and Bohr's model, Dellsén is ignoring this fact. Taking the explanation of the stability of atoms provided by Bohr's model into account demonstrates how the transition from Rutherford's model to Bohr's model increased understanding of the atom. Rutherford's model could not explain the stability of atoms, but Bohr's model did. Sure, Bohr's model was not without issues, either. While the atomic structure suggested by the model explained the stability of atoms, the structure and features that Bohr postulated could not be as straightforwardly explained. Yet, these stipulations could at least be justified by referring to other phenomena like the photoelectric effect and early quantum theory that are in accordance with the model, which is exactly what Bohr himself did. Again, as I argued in the case of explanatory brute facts as well, understanding and explaining a phenomenon does not entail the understanding and explanation of all the information that is involved in the understanding and explanation of the phenomenon.

Even if my claim that the stability of the atom was the more important aspect for Bohr than the discrete wavelengths of spectral lines emitted by atoms is wrong, one could still question whether the atom and the emission of spectral lines at certain wavelengths are as distinct phenomena as Dellsén suggests. If scientists want to understand a phenomenon (the atom in this case), they will want to understand every feature of this phenomenon. Since the emission of spectral lines of certain wavelengths is a feature of atoms, to understand the atom in its entirety, the emission of spectral lines must be understood as well. Likely, physicists would not claim that they fully understand the atom if they have no clue why or how atoms emit spectral lines at the wavelength at which they do. Understanding comes in degrees, Dellsén and I agree on this. Therefore, gaining understanding of a phenomenon usually takes time, as its manifestation process is iterative. Grasping and explaining some relations will enable grasping and explaining further relations. Even if one argues that physicists did not explain (or understand) the atom through Bohr's model, yet, but merely the emission of spectral lines, it would be strange to claim that they understood a completely different or unrelated phenomenon. They understood a feature, the emission of spectral line, of the phenomenon, the atom, they ultimately wanted to understand. Bohr's model did not provide an ultimate explanation of the atom, but it enabled new research routes for physicists and pathed the way for the development of the valence shell model which is used today. Understanding the emission of spectral lines can be taken as one step in the process of understanding the target phenomenon, the atom.

#### 6.2.3.3 Galileo's thought experiment, again

The third and last type of cases, which his supposed to be covered by DMA but not by *U→E*, is labelled 'explanatory disconnectedness' by Dellsén. He illustrates this type with Galileo's thought experiment introduced into the debate on understanding by Peter Lipton. I discuss this example and Lipton's view in general in section 3.2. As a memory aid, Dellsén presents this example in the following way.

> The reductio is a thought experiment in which we suppose that a lighter object is fastened to a heavier object. If lighter objects accelerate slower, then the lighter object should slow down the heavier object, so the two objects should accelerate slower than the heavier object would by itself. However, the two objects can also be considered together as one larger object, which is thus heavier than either of the objects that it is composed of, so this composite object should accelerate faster than the heavier object. But since the two objects cannot both accelerate faster and slower than the heavier object would by itself, the idea that heavier objects accelerate faster than lighter objects cannot be correct.[72]

Dellsén agrees with Lipton that the thought experiment provides understanding, but his analysis differs significantly. According to Lipton, the thought experiment, while not providing an explanation, displays a necessity. It shows that gravitational acceleration must be independent of mass.

> However, [Dellsén] fail[s] to see how the necessity of the fact that gravitational acceleration is independent of mass is responsible for our understanding in this case. In [his] view, Galileo's reductio instead shows that understanding can be increased by grasping that two factors are independent, whether by necessity or as a contingent matter. In other words, Galileo's reduction provides understanding not by showing necessity, but by showing a certain kind of independence.[73]

That is, understanding increases when we become aware that two seemingly related factors are actually independent from one another. Dellsén's DMA can nicely capture these cases, since understanding can increase either through improving the accuracy or the comprehensiveness of the dependency model of the phenomenon. In the example of gravitational acceleration, Galileo's thought experiment increases

72 Dellsén (2020), p. 1278.

73 Ibid. p. 1279.

the comprehensiveness of the dependency model by showing that gravitational acceleration and mass are independent.[74]

> Of course, this increase in understanding is by itself rather modest according to DMA, since it does not tell us what factors gravitational acceleration does depend on, only that a particular contextually salient factor, namely, mass, is not one of these. Again, this appears to be a correct prediction, since the understanding of gravitational acceleration provided by Galileo's reductio is indeed rather incomplete.[75]

As in the two previous examples, I disagree that understanding of gravitational acceleration increases, changes or becomes possible at all without explanation. Dellsén argues for the Galileo example that grasping the independence of gravitational acceleration from the mass of falling bodies enables understanding of gravitational acceleration as independent of mass. This is correct, but not the whole story. The crucial question is what is required to realize that two factors are independent. Why should we accept the independence of two factors that so far seem to be dependent? We should accept the independence of two factors if we have reasons for doing so. The Galilean thought experiment did not only show that gravitational acceleration is independent of mass, but additionally provided an explanans, a reason, why this is the case, namely because it is logically impossible that gravitational acceleration is dependent on the mass. Arguing on the basis of the logical impossibility that gravitational acceleration is independent of mass is more than merely finding out that gravitational acceleration is independent of mass and not having any reason or explanans to make sense of that fact. This is the additional component that Dellsén is missing.

In general, why is Dellsén's DMA insufficient for an account of understanding? Because grasping dependence relations of a phenomenon is not sufficient for understanding. For Dellsén, understanding is "roughly the possession of a model of the understood phenomenon's dependence relations."[76] I am sympathetic to that view, since it is compatible with my requirement that relations of a phenomenon need to be grasped if the phenomenon should be understood. I am not claiming that DMA is fundamentally wrong. I am claiming that it is incomplete for understanding. The crucial point is that we need to provide reasons why we think that our dependency model increased in accuracy or comprehensiveness. If we cannot provide reasons for the improved accuracy or comprehensiveness of a dependency model, how could we know that it improved at all? And providing reasons why an aspect of a phenomenon

---

74 See ibid. p. 1279.

75 Ibid. pp. 1279f.

76 Ibid. p. 1280.

or its dependency model is taken to be like this or that is everything I require from my generic conception of explanation introduced in section 3.1.

The deficiency of Dellsén's account becomes clearer when we apply it to the case of the research on the zebrafish *oep*-mutant, which I present in chapter 5.1. One observation that the biologists wanted to understand was the significant similarity of the *oep*-mutant phenotype and the *cyc/sqt*-mutant phenotype. After observing the similarity, the biologists had the idea that Oep and Nodal, the proteins that are missing in one of the two mutant strains, respectively, may act in a common pathway. This hypothesis would explain why both mutant strains look similar, since in both cases one component of the pathway is missing and therefore, the pathway would not function properly in either of the two mutant strains. In a next step, the biologists tested this hypothesis by injecting mRNA's encoding Nodal in the *oep*-mutants, which should replace the function of Oep. The biologists wanted to test the possibility that their hypothesis is wrong and that Oep and Nodal do not act in a common pathway. Although the experiment confirmed the hypothesis that Oep is indeed essential for Nodal signaling, let us consider the counterfactual case, that the experiment would have shown that Nodal signaling takes place without Oep, that the presence and function of both proteins are or can be independent from each other. If this had been the result of the experiment, the scientists would have known of their independence and could have explained why the proteins are independent on the basis of their experiment and results, but they would not have understood the similarity of the phenotypes of the two mutant strains. They would have had no clue why the two mutants have a similar phenotype. Again, as in the two other cases before, if DMA designates any type of understanding in cases of explanatory disconnectedness, it will not be the type of understanding that scientists want to have. Therefore, Dellsén's account is at least insufficient for scientific understanding.

#### 6.2.3.4 Understanding requires explanation

In sum, Dellsén's DMA and the GE-account of understanding agree in one crucial aspect, while disagreeing fundamentally on another. His DMA is compatible with the GE-account in so far as we both take grasping of (dependence) relations to be crucial for understanding. Although Dellsén speaks of grasping a *dependency model* of the phenomenon while I require grasping *relations* of the phenomenon, I do not think that this conceptual difference is as substantial as it may seem. Since Dellsén argues that models involved in understanding aim to capture the network of dependence relations that a phenomenon stands in, I do not see a disagreement with my claim that a scientist needs to grasp relations of the phenomenon without them being necessarily mediated by a model. In Dellsén's view, models are information structures of some kind that are interpreted so as to represent their targets. Whether models conceptualized in this way are necessary for understanding phenomena or whether it is possible to grasp relations of a phenomenon without such kinds of models or any other kind of mediator remains a question for further research. I am not denying that the GE-account may lack an important aspect here, since I do not analyze the function of models for understanding. Nonetheless, the crucial agreement of DMA and the GE-account is that (dependence) relations of phenomena need to be grasped by a subject.

The decisive disagreement between Dellsén's DMA and the GE-account concerns the role of explanation for understanding. While I take explanation to be necessary for understanding, as de Regt and Khalifa, Dellsén wants to show with his DMA that understanding does not require explanation. As I have shown in the discussion of the three examples provided by Dellsén, he fails to make a convincing point that (scientific) understanding is possible without explanation. In the case of explanatorily brute or not yet explained facts, I do not see in what sense scientists have understanding of these facts. Actually, the attempts of physicists to find explanations of the values of dimensionless physical constants rather suggests that physicists do not understand the values of these constants, yet. In the case of Bohr's model of the atom as an instance of failed explanatory targetedness, Dellsén ignores the successful use of the model to explain the stability of atoms, in addition to explaining the emission of spectral lines at certain wavelengths. For the third type of cases exemplified with Galileo's thought experiment, it is also not clear to me what exactly the understanding consists in. If I come to realize that two factors A and B are not related in the way I thought they are, I may realize that I misunderstood the relation of A and B. But without getting any explanation of why A and B are not related in this way or why they are not related at all, I do not replace my misunderstanding with understanding of the relation. Instead, I replace my misunderstanding of the relation of A and B with no understanding at all.

It is worth noting that Dellsén, although he is arguing for a type of objectual understanding, does take understanding and explanation to be very closely related. He explicitly admits that "explanatory accounts of understanding can seem plausible, perhaps even irresistible, because understanding does tend to bring increased capacities to explain. In that sense, explanation and understanding are indeed closely linked."[77]

> [Furthermore,] although [the DMA] account is not an explanatory account of understanding, it does preserve the kernel of truth in explanatory accounts in so far as a sufficiently accurate and comprehensive dependency model contains the sort of information about a phenomenon that is required to explain it and related phenomena, provided that they can be explained at all. This is so for the simple reason that the dependence relations that these models must correctly represent in or-

77 Ibid. p. 1277.

der to provide understanding (for example, causal and grounding relations) are precisely the sort of relations that form the basis for correct explanations.[78]

Given these confessions, it is quite surprising that Dellsén writes a whole paper on understanding without explanation. I suspect the crux lies in the standard conception of explanatory accounts of understanding that Dellsén is using, according to which understanding stems from grasping, knowing, or having an explanation. That is, explanation comes first, understanding second. As I argue at length in section 6.1 and in chapter four, I do not consent to this view and turn the order around. The ability to understand comes first and with the help of available knowledge, equipment and further skills, a new explanation comes second through the manifestation of the ability to understand.

## 6.3 Understanding is an impressive cognitive achievement and a goal of science

What is scientific understanding and how is it achieved? I have presented and elaborated the GE-account of scientific understanding to answer these questions. According to the GE-account, a scientist has scientific understanding if and only if the scientist is able grasp relations a phenomenon stands in and to articulate these relations in form of new explanations of (aspects of) the phenomenon. Understanding is an ability that is manifested through the iterative processes of grasping some relations and articulating (hypothetical) explanations, and improved through grasping more (aspects of) relations and confirming or revising explanations. For grasping relations and articulating them in explanations, the scientist has to possess and use necessary equipment, relevant knowledge and research skills. Additionally, she has to be a member of a scientific community. The community is a decisive contextual factor for understanding, as it provides its members with the necessary resources, including knowledge, skills, and further equipment, that enable scientists to understand a phenomenon. Moreover, young scientists acquire and train the ability to scientifically understand phenomena in the first place through engaging with more experienced members of their community, through guidance by their professors and supervisors. Additionally, to ensure that the understanding gained by an individual scientist counts as scientific understanding and as objective, and not as some form of non-scientific or inappropriate understanding, parts of the scientific community need to assess and approve the understanding gained by individuals. By making the individual understanding publicly accessible, its objectivity increases and its status as scientific can be confirmed.

78 Ibid. pp. 1282f.

I have also compared the GE-account to other accounts of understanding to highlight its merits. Despite the agreements with de Regt's account, Khalifa's EKS model, and Dellsén's DMA, the GE-account does diverge from these views in various respects. In contrast to de Regt, theories do not take a center stage in the GE-account, which makes it possible for the GE-account to accommodate cases from scientific practice in which scientific theories are not (yet) available or do not play a decisive role in the manifestation of understanding. Furthermore, de Regt primarily analyzes understanding of theories, and not understanding of phenomena, although the later one is taken to be the main aim of science. I agree with the basic distinction and that it is necessary to first understand a theory if one wants to understand a phenomenon through that theory. However, my target of analysis is the understanding of phenomena as an ultimate aim of science. In that sense, the GE-account can be seen as an extension of or contemplation to de Regt's account. The disagreement with Khalifa is more fundamental, as he takes understanding to be kind of knowledge and a product of scientific research, while I view the iterative processes of grasping relations and articulating explanations to be the manifestation of understanding through which new knowledge, alias an explanation, is produced. These two basic intuitions may be incompatible and an agreement might never be reached. Still, I take an ability-account of understanding to be more suitable to capture what we want to see if we test one's understanding. In such cases, we do not merely want someone repeating or rephrasing known explanations. Rather, we expect that this person is capable of using and applying available knowledge in a novel way that cannot be prescribed beforehand. And concerning Dellsén's DMA, I disagree that grasping (dependence) relations of a phenomenon is sufficient for understanding that phenomenon. This is so because grasping, conceptualized as having a relation between one's mind and world, as having epistemic access to a worldly phenomenon, which is the conception that Dellsén and I adopt, is not enough for making sense of a phenomenon. Grasping a (dependence) relation is the first necessary step, but merely having access to features of a phenomenon is not identical to figuring out why a phenomenon has precisely these features or how exactly they are related to other features of the phenomenon. Without researching grasped relations further and articulating the acquired insights in form of explanations, scientists will not understand the features of the phenomenon they grasped, as they could not make sense of them. Hence, understanding phenomena scientifically is an ability, manifested in grasping relations and articulating explanations, that exceeds any account of propositional knowledge and that does not necessarily require theories.

Despite all the disputes and conflicting positions concerning the nature, conditions, and various characteristics of understanding, there is also a common ground shared by all scholars engaged with the topic, which should not be forgotten or ignored. Everyone agrees that understanding is an impressive cognitive achievement that comes in degrees and is impacted by the context in which it is achieved. Furthermore, understanding is a goal of every epistemic endeavor, especially but not exclusively for science. Everyone, scientists and non-scientists, strive for understanding something in some domain. Hence, keeping this common ground in mind will enable and empower future research and insights on understanding.

# 7. Conclusion

Understanding is an intellectual achievement to which human beings permanently aspire, in their everyday and professional lives. The understanding that scientists want to achieve of the empirical phenomena they investigate in their research is no exception. Gaining understanding of phenomena is a central epistemic aim of science, as is the understanding of phenomena, experiences, or situations in other domains of human life. This book is targeted at scientific understanding, the understanding that scientists *qua* scientists achieve of the phenomena they are researching. It provides a novel account of scientific understanding that answers the questions what scientific understanding is and how scientists achieve it. While it was not the goal of this book to provide an account of understanding in general, i.e. understanding that any human agent can gain of anything in any context, the insights about scientific understanding presented in this book will be meaningful for the investigation of understanding in general, and of scientific understanding in particular. In this final chapter, section 7.1 provides a summary of the arguments and results obtained in this book, before I present an outlook on follow up questions concerning (scientific) understanding future research could address.

## 7.1 Summary

This book was driven by two main question: what is scientific understanding and how do scientists achieve it? I developed an account of scientific understanding that answers these questions by providing necessary and sufficient conditions for understanding. Let me recap how I arrived at my account.

I started with a survey of the current philosophical literature on scientific understanding and presented three elaborate and prominent accounts of scientific understanding provided by Henk de Regt, Kareem Khalifa, and Finnur Dellsén in chapter two. The comparison of these positions revealed common ground, as well as disagreements between these scholars. Regarding the common ground, I identified four shared intuitions or assumptions: first, the topic of interest is that of understanding gained in science in general, and not in specific scientific disciplines. Second, the author's accounts focus on understanding that individual scientists can achieve, and not on a form of collective understanding that a group or bigger community could gain. Third, it is agreed that understanding of worldly phenomena is an ultimate aim of science, and hence more important and interesting for philosophical analyses than understanding of theories or models used to achieve understanding of phenomena. And last but not least, while all three scholars formulate accounts of understanding gained in science in general, they are all aware that contextual factors, such as specific historical or disciplinary circumstances or local constraints, have an impact on understanding. Therefore, any account of scientific understanding must leave room for contextual variation. I adopted these assumptions as the basis for developing the "Grasping and Explaining"-account of scientific understanding, or GE-account for short.

Regarding the disagreements among de Regt, Khalifa and Dellsén, I detected two central controversial questions concerning scientific understanding:

1) Does scientific understanding require explanation or not?
2) Is understanding an ability or a type of knowledge?

These questions needed to be addressed for an account of scientific understanding to be provided. Hence, I first turned to these two questions identified in chapter two, before I actually developed answers to the two main questions of this book.

Chapter three engaged with the relation of scientific understanding and scientific explanation. I presented and discussed several positions according to which understanding and explanation can be distinct, that understanding does not (always) require explanation, and several counterarguments. Following a clarification of my conception of *explanation* in section 3.1, I engaged with Peter Lipton's view in section 3.2. Lipton defends a separation of understanding and explanation and argues for this by using four examples in which understanding and explanation apparently fall apart. I argued that Lipton fails to show that understanding is achieved without explanation in his examples. Section 3.3 was devoted to Jonathan Kvanvig's influential differentiation between objectual and explanatory understanding, and Kareem Khalifa's reductionist counterarguments against Kvanvig's conception of this differentiation. After a consideration of additional arguments in favor of as well as against a differentiation of objectual and explanatory understanding in section 3.4, I concluded in section 3.5 that scientific understanding does require explanation, that a differentiation between objectual and explanatory understanding is not feasible in the case of scientific understanding.

This is the case because, first, all proponents of a separation of understanding and explanation employ a very narrow notion of explanation that is restricted to causal explanation. Since extensive work on scientific explanations done by philosophers of science revealed the legitimate presence and use of various different types of explanations across the sciences, an explanatory monism, i.e. any restriction to one specific type of explanation, e.g. causal explanation, is unjustified. In contrast, a pluralist position concerning scientific explanation should be adopted. Second, and related to the previous point, explanations are omnipresent in and an undeniable goal of science. Therefore, it is much more plausible to conceive scientific understanding as requiring scientific explanation, since such a conception does naturally relate two aims of science, achieving understanding and achieving explanation. Tearing scientific understanding and scientific explanation apart is an implausible move in light of these two goals and the ubiquity of explanation in science.

Chapter four addressed the question of whether scientific understanding is an ability or a type of (propositional) knowledge. These are the two options currently discussed in the philosophical debate on understanding. I consent to the first view and take understanding to be an ability, for which position I argued in this fourth chapter. I started with a clarification of the notion *ability* in section 4.1. After having explained what I mean by *abilities*, namely dispositions to succeed, I used section 4.2 to argue that understanding should plausibly be conceived as an ability, and not as a type of (propositional) knowledge. I claimed that understanding is the ability to make sense of a phenomenon or other entities that someone wants to understand, e.g. experiences, situations, theories, poems and so on. If understanding is an ability, a disposition to succeed, it needs to be manifested somehow.

Hence, section 4.3 was devoted to the manifestation of the ability to understand something. Based on discussions of the very prominent notion of grasping within the philosophical debate on understanding and my demand that scientific understanding requires explanation, I argued that scientific understanding is manifested in the process of grasping relations the object of understanding stands in (may it be a phenomenon, situation, experience, theory, or poem) and in articulating the grasped relations in form of explanations. In the first part of the manifestation process, through grasping, an agent establishes some connection between her mind and the thing in the world she wants to understand. Through grasping, an agent recognizes, becomes aware of, or "sees" some relation the thing stands in. And in the second part, by articulating an explanation, she applies knowledge and concepts she already possesses to the relation she grasped in order to clarify or work out what exactly she grasped, what kind of relation it is and what the relata are.

So, understanding, or making sense of, something manifests in the process of grasping some relation of the thing that shall be understood and in sorting out what precisely is grasped through articulating the grasped relation in an explanation. This conceptualization of the manifestation of understanding does not only accommodate the intuitions most people (including philosophers) have when thinking about understanding, namely that understanding is something like 'seeing how things hang together', it also resolves conflicting and confusing ideas about the nature and relation of understanding towards knowledge and explanation. Understanding and explanation cannot be torn apart, since explanations are the products of the manifestation of understanding. And understanding and knowledge are inextricably intertwined, as one cannot understand anything in the world without resorting to some already existing knowledge, and new (explanatory) knowledge is also produced via understanding.

Then, I presented an episode from scientific practice and my philosophical analysis of it in chapter five. This episode was about the introduction of a new model organism into biological research, the zebrafish, and how this model organism enabled scientific understanding of the genetic regulation of vertebrate development. In section 5.1, I first depicted the episode from research practice in biology by describing the historical context, aims, challenges, and developments that ultimately resulted in the establishment of zebrafish as a new model organism, the emergence of a new research discipline, and new insights into genetic regulations underlying vertebrate development. Following the historical overview, in section 5.2 I analyzed how exactly the scientists involved in this research episode gained understanding of the phenomenon that was the target of their investigations, the genetic regulation of vertebrate development, as well as which contextual factors had an impact on the understanding or enabled understanding at all. I argued that the episode from scientific practice does not only second my claims devised in the previous two chapters, that understanding is an ability that requires explanation, but also brings three additional insights about scientific understanding to the fore.

First, scientists needed specific knowledge, research skills, and equipment for understanding the specific phenomenon they were interested in. Second, in order to gain all these necessary resources, scientists had to establish an appropriate infrastructure or community that could provide all its members with these resources. And third, the scientific episode spotlighted the iterative nature of the manifestation process of understanding. This means that scientific understanding is (usually) not manifested in a two-step process of first grasping and then explaining, but rather in several subsequent steps or instances of grasping some relation or aspects of a relation, articulating the grasped aspect in an explanation, which enables grasping of further aspects of this relation or an additional relation, which is articulated in an explanation again, and so on and so forth. That is, the manifestation process of scientific understanding is much more demanding and complex than it appeared given my argumentation in chapter four. Thus, the episode of biological practice can be deemed useful in providing important and novel insights for a philosophical analysis of scientific understanding.

Having everything we need, chapter six finally provided the space for the account of scientific understanding I developed based on the work done in the previous chapters. This account, the "Grasping and Explaining"-account of scientific understanding (or GE-account for short), as I have termed it, is my answer to the main questions I want to answer with this book: what is scientific understanding and how do scientists achieve it? The GE-account takes the following form:

A scientist *S* has scientific understanding of an empirical phenomenon *P* in a context *C* if and only if

i.) *S* grasps (details of) relations that *P* stands in and articulates these relations in the form of new explanations of (aspects of) *P* (*manifestation condition*),
ii.) *S* possesses and uses (material) equipment, relevant knowledge and research skills provided by *C* and required for understanding *P* (*resource condition*), and
iii.) *S* is a member of a scientific community that enables *S* to understand *P* and parts of that community approve *S*'s understanding of *P* (*justification condition*).

In section 6.1, I elaborated on the scope and the three conditions of the GE-account, which I called *manifestation condition*, *resource condition*, and *justification condition* and take to be necessary and sufficient for scientific understanding. In a nutshell, the GE-account only captures scientific understanding of an empirical phenomenon gained by an individual scientist who is situated in a specific context, which impacts the understanding in several ways that are spelled out in the three conditions. The *manifestation condition*, as its name already suggests, expresses the manifestation process of understanding, namely grasping relations and articulating explanations. The necessary resources that a scientist needs if she wants to manifest her understanding of some phenomenon are covered by the second condition, the *resource condition*. Finally, the *justification condition* gives the respective scientific community of a researcher its proper due, as it is the scientific community that, first, provides any researcher with the resources necessary to do research in the respective field at all, and second, assesses and eventually approves the understanding that researchers gain. In other words, no scientist is justified in thinking that she understood some feature of a phenomenon if no other members of her community accept the explanation articulated by her, and hence her ability to understand the phenomenon, as legitimate or appropriate.

Finally, in section 6.2, I demarcated the GE-account of scientific understanding from the other accounts introduced in chapter two and highlighted its advantages in comparison to these other accounts. Since scientific theories do not play a decisive role in the GE-account, it has a greater flexibility and can better accommodate cases from scientific practice in which theories were either completely absent or had no crucial function for the understanding. This is an advantage of the GE-account in comparison to de Regt's account of scientific understanding, according to which phenomena cannot be understood without theories. Contrary to Khalifa, who defines understanding as knowledge of explanations, I argued that any account that conceptualizes understanding as an ability and not as knowledge, like the GE-account, can better capture what we intuitively expect from someone who has understanding of some phenomenon. That is, we expect such persons to be able to somehow engage with the phenomenon, to work with or on it, to generate novel insights about it, and the like. None of these activities can be captured by accounts that view understanding as a type of (propositional) knowledge. Lastly, I argued against Dellsén that it is implausible to conceptualize scientific understanding as not requiring or being independent from explanation. Since Dellsén and I employ the same conception of grasping, I held that grasping is not sufficient for understanding, as grasping does not comprise the process of working out what exactly has been grasped. Understanding or making sense of phenomena requires the application of available knowledge or concepts to the features of the world that were grasped, and this second step, the articulation of an explanation, exceeds grasping.

## 7.2 Outlook

While this book hopefully provides answers to the questions what scientific understanding is and how scientists achieve it, there are, of course, many unresolved issues that arise around understanding in general and scientific understanding in particular. I will provide a short outlook on some questions that derive from the GE-account developed in this book.

### 7.2.1 Understanding and representation

First, the GE-account only captures understanding that scientists gain of the phenomena they are researching. It does not address the question of what it means for scientists to understand the various representations that are used in research in order to achieve the ultimate goal, understanding of phenomena. As I stated in the introduction to this book, the two major issues that are of interest to philosophers of science, at least according to Stephen Grimm, are the relations of understanding to explanation and to idealizations or models.[1] While I intensively worked on the relation of understanding and explanation, I did not engage with the relation of understanding to other types of models used in science.

Representations are very diverse, ranging from explanations and theories to classificatory systems like the periodic table, graphical depictions, to various kinds of models, abstract or concrete ones, and to computer simulations and artificial intelligence systems. While there is widespread agreement that, in most cases, scientists have to understand the representations through or with which they

1 See Grimm (2021).

understand some phenomenon, it is not at all clear how the understanding of representations relates to the understanding of real world phenomena, especially when complex computer simulations or AI systems are involved.[2] In short, one question is whether the understanding of some phenomenon necessarily requires the understanding of the representations used for understanding this phenomenon. And if this question is answered in the affirmative, a follow-up question will be whether there are any differences between understanding a representation on the one hand, and understanding some phenomenon through or with that representation on the other.

These questions are already intensively addressed in the debate on understanding. If it is the case that understanding some or specific representations is necessary in order to understand phenomena (for instance, that a physicist must understand electromagnetic theory based on the Maxwell equations if she wants to understand electromagnetic phenomena), is understanding these representations in any way different from understanding the phenomenon? And how might understanding a representation then be related to understanding a phenomenon? Henk de Regt provides one answer to these questions, as he conceptualizes understanding of phenomena (UP) and understanding of theories (UT) differently. I presented his view in section 2.1. In a nutshell, de Regt argues that UT and UP are necessarily intertwined. Scientist need to understand a theory in order to construct models, and hence explanations, of phenomena, and thereby understand said phenomena.[3] In the examples that de Regt discusses, scientists had to understand a theory first, before they could use the respective theory to construct explanations and understand a phenomenon.

In contrast, the GE-account does not address potential differences between understanding phenomena and understanding theories or other representations in science, and hence does not analyze the possible relations between these types of understanding (assuming that they can reasonably be viewed as two different types). While I argued for the advantage of this characteristic of the GE-account in section 6.2.1, namely that the GE-account can accommodate instances of scientific understanding of phenomena that were possible without drawing any specific theory or in which an involved theory did not deliver the crucial insights, theories are of course of crucial importance in many fields of scientific research. It might be interesting

2 For investigations on whether and how AI systems or artificial neural networks could or must (not) be understood in order to use them for understanding empirical phenomena, see for instance Sullivan, E. (2020), "Understanding from machine learning models." *The British Journal for the Philosophy of Science*, DOI: 10.1093/bjps/axz035; or Rudin, C., et al. (2021). "Interpretable machine learning: Fundamental principles and 10 grand challenges." *Statistics Surveys*, 16, pp. 1–85, DOI: 10.1214/21-SS133; or Chirimuuta, M. (2021). "Prediction versus understanding in computationally enhanced neuroscience." *Synthese*, 199 (1), pp. 767–790, DOI: 10.1007/s11229-020-02713-0.

3 See de Regt (2017), especially chapters two and four.

to analyze under which conditions and how exactly theories might be indispensable for achieving scientific understanding of phenomena.

Explanation is another central notion in science with which I extensively engaged in this project. According to the GE-account, a phenomenon is understood through grasping relations and articulating explanations. In other words, scientists understand a phenomenon for which there is no explanation through articulating one. Is there an intrinsic difference between understanding conceptualized like this, and instances in which some phenomenon is understood through an already available theory or explanation? If a theory or explanation is already available, will the ability to understand a phenomenon be manifested differently than in cases in which a theory or explanation is articulated during the manifestation process of understanding? For example, could it be that in instances in which an explanation is already available, the manifestation of understanding only comprises the grasping of relations represented by the explanation? If this is the case, what exactly is understood?

If a subject grasps relations presented by an explanation or theory, does she then understand the phenomenon that is represented, or does she merely understand the explanation or theory itself? Mark Newman, whose view I presented in section 4.3.2, distinguishes between three different types of understanding: knowing an explanation (i.e. understanding an explanation linguistically), understanding an explanation (having explanatory understanding of the phenomenon represented by the explanation) and theoretical understanding (understanding a theory).[4] Can these three types of understanding plausibly be separated? Is there a difference between understanding a theory or an explanation, understanding a phenomenon through a theory or explanation that is already available, and understanding a phenomenon through constructing a theory or explanation, which is the kind of understanding the GE-account captures?

These and similar questions become even more pressing if we do not only consider theories or explanations, which are often conceived of as having some propositional or explicit mathematical form, but other kinds of representations in science, especially models. While some scholars identify models with explanations or theories, there undeniably is a wide variety of models used in science of which one could ask whether these models have different functions for understanding or whether the understanding of phenomena varies in some sense when different kinds of models are used. How different kinds of models can be differentiated is another challenging question. One potential classification is provided by Weisberg, who distinguishes material or concrete models from theoretical or abstract models, as well as from computer models.[5] The Crick and Watson model of DNA that is built with real

4 See Newman (2017).

5 See Weisberg (2013), especially chapter two.

physical balls and sticks is a material model. The Lotka-Volterra model of predator-prey dynamics takes the form of four differential equations and is, hence, a theoretical model. And then there are computer models used for running simulations in, for example, climate science or epidemiology. Again, we can ask, and a considerable number of scholars already has,[6] whether scientists who use any such models must understand the model if they want to understand a phenomenon using the model, what exactly understanding any of these kinds of models amounts to, and whether this understanding is in some sense different from understanding a phenomenon with these models.

Originally, I wanted to address these kinds of questions in my project. My idea was to analyze two different scientific episodes, one on the use of model organisms in biology, and the other on research with computer models in climate science. Investigating the use of these two different kinds of models in the respective disciplines might have revealed significant differences in the understanding through material models versus the understanding by using computer models. Alternatively, the analyses would have shown that there is no significant difference. While I do not know the answer to this question, as I could not conduct the comparative analysis due to time constraints, my assumption is that there might be a difference. The reason for this assumptions is that in the case of model organisms, which can be considered as a special sub-type of material models, scientists directly intervene in the mechanism, i.e. the phenomenon they want to understand, as I explain in section 5.1. In studies of zebrafish, biologists directly manipulated biochemical pathways which they wanted to understand. Once identified and manipulated, they inferred inductively that the specific gene in question, or its orthologs, have the same function in the embryonic development of other species.

Nothing like this happens in climate science. While one could claim that humanity has been running one major experiment with our climate for decades, this is not what climate scientists do in their attempts to understand the mechanisms of the earth's climate. Rather, climate scientists use computer models to run simulations of the global climate, and through these simulation runs they gain information about the model world defined by the parameters used in the model. These insights need to be related to real world-climate, as any computer model is in some sense an inadequate representation of the real phenomenon. In a nutshell, climate scientists do not directly intervene in the world's climate, but merely in model worlds inscribed in the computer models. Whatever climate scientists learn about the model world

6 See for example Elgin (2007); Elgin (2017); de Regt, H. & Gijsbers, V. (2017), "How False Theories Can Yield Genuine Understanding." In Grimm, S., Baumberger, C. & Ammon, S. (eds.), *Explaining Understanding. New Perspectives from Epistemology and Philosophy of Science*, pp. 50–75, New York and London, Routledge; Reutlinger, Hangleiter & Hartmann (2018); or Strevens (2017).

needs to be translated to the real world. Thus, it seems that model organisms allow for a much more direct access to the phenomenon in question, while computer models can only provide indirect access that requires some additional step.

This additional step could be interpretation. Is interpretation just another term for understanding or are interpretation and understanding two different things? Could it make sense to argue that scientists need to *interpret* a theory or a model in order to *understand* a phenomenon? For Michael Polanyi, whose views on tacit knowledge and understanding I extensively used in chapter four, there is no fundamental difference between these notions. According to him, understanding and interpretation are basically one and the same thing, the only difference being that interpretation requires language.

Certain animals and very young children are able to understand things happening in the world without possessing or being able to use any articulate language. Polanyi claims that (some) animals and young children can gain understanding, but not interpretation, as he takes interpretation to be a more sophisticated type of understanding.[7] Analyzing the concept of interpretation in relation to understanding might be helpful in clarifying ideas about understanding representations and understanding the phenomena they represent, since interpretation is a crucial concept in the philosophical literature on scientific representation. For instance, Richard I. G. Hughes suggested his DDI account of scientific representation (denotation, demonstration, and interpretation), without really explicating what he means by interpretation, unfortunately.[8] His main idea remained influential nevertheless. Gabriele Contessa, for example, offered his interpretational account of scientific representation including a neat characterization of what he means by interpretation within his account.[9] In sum, my assumption is that the rich literature on scientific representation might provide important and relevant insights for analyzing the nature of understanding representations, its relation to understanding phenomena that are represented, as well as the notion of interpretation in this context.[10]

To conclude, I did not engage with the function of theories, models, or other types of representation for scientific understanding of phenomena in the course of my project. Therefore, the GE-account of scientific understanding does not capture the understanding that scientists might have of the theories or models they employ

7 See Polanyi (1962 [1958]), especially chapter five.

8 See Hughes, R. I.G. (1997), "Models and Representation", *Philosophy of Science*, 64, pp. 325–336, DOI: 10.1086/392611.

9 See Contessa, G. (2007), "Scientific Representation, Interpretation, and Surrogative Reasoning", *Philosophy of Science*, 74 (1), pp. 48–68, DOI: 10.1086/519478.

10 For an overview on various accounts of scientific representation, see for example Frigg & Nguyen (2021).

in their research. However, since theories and various types of models are undoubtedly extensively used in science, their relation to, and function for, understanding phenomena should be taken into account and deserves further philosophical analysis.

### 7.2.2 Understanding and prediction

As I have claimed throughout this book, understanding and explanation are two central and interrelated goals of science. This view is widely shared and not seriously contested to my knowledge. However, one might be missing another central goal of science: prediction. I did not engage with the relation of understanding to prediction in this project, but I do think that this is a very important question. Hence, I would like to at least point towards discussing this issue.

With whom should I start, if not with the founding father of the philosophical debate on understanding? Henk de Regt also addressed the relation of understanding to prediction, although not as detailed as the relation of understanding to explanation. The notion of prediction sneaks into de Regt's account of understanding via his criterion of the intelligibility of theories. Again, according to de Regt, scientists can understand phenomena only through the understanding of theories, and specific scientists in specific contexts have understanding of a particular theory if that theory is intelligible to them. While de Regt admits that different criteria might be employed to determine the intelligibility of theories in different historical or disciplinary contexts, he proposes and discusses one specific Criterion for the Intelligibility of Theories, which he takes to be especially suitable to accommodate the physical sciences:

> **$CIT_1$:** A scientific theory *T* (in one or more of its representations) is intelligible for scientists (in context C) if they can recognize qualitatively characteristic consequences of *T* without performing exact calculations.[11]

De Regt demands that if a theory is intelligible for scientists, the scientists will be able to make rough qualitative predictions that turn out to be correct to some degree when tested. Successful predictions allow for the construction and testing of explanations, and hence understanding, of phenomena. And some degree of understanding of phenomena, in turn, will enable new successful predictions. Therefore, according to de Regt, explanation, understanding, and prediction are interrelated goals of science and cannot do without each other:

---

11 De Regt (2017), p. 102.

> Compare a successful scientific theory with a hypothetical oracle whose pronouncements always prove true. In the latter case, empirical adequacy would be ensued, but we would not speak of a great scientific success (and perhaps not even of science *tout court*) because there is no understanding of how the perfect predictions were brought about. An oracle is nothing but a black box that produces seemingly arbitrary predictions. Scientists want more than this: in addition they want insights, and therefore need to open the black box and consider the workings of the theory that generates the predictions.[12]

Such a view on the interconnectedness of explanation, understanding, and prediction poses great challenges to branches of research in which some kinds of black box models, e.g. machine learning models, are used. However, de Regt's position is criticized, for example by Johannes Findl & Javier Suárez.[13]

Findl & Suárez argue that one can gain understanding of phenomena through purely statistical models without any causal knowledge, as these models provide predictions. Hence, the authors differentiate between predictive understanding, as they call it, and explanatory understanding and argue that understanding through prediction and without explanation is possible. The basis for this claim by Findl & Suárez is a case study on the use of epidemiological models in the COVID-19 pandemic:

> Early versions of such models based their predictions on statistical data that had been provided by other countries, rather than on a causal understanding of the disease. In other words, early COVID-19 models were what epidemiologists call *statistical models*, i.e., models that derive their estimations from a regression analysis that fits a curve to empirical data — such as the number of infections or deaths — rather than from causal data about the patterns of infection of the disease which were mostly unknown at the time. [...] While these [purely predictive] models did not include specific causal-mechanistic information about how the disease would spread or affect those infected, their primary function was to give estimates of what would most likely happen if counter-measures were introduced or removed.[14]

These statistical models were continuously modified and updated on the basis of newly available data from countermeasures and their effects in specific geographical regions. If predictions yielded by a model did not fit empirical data about, for example, the infection rate, incorrect or missing assumptions in the model had to be

12 Ibid. pp. 101f.

13 See Findl, J. & Suárez, J. (2021), "Descriptive understanding and prediction in COVID-19 modelling." *History and Philosophy of the Life Sciences*, 43 (4), pp. 1–31, DOI: 10.1007/s40656-021-00461-z.

14 Ibid. p. 3, original emphasis.

corrected or added. This procedure improved the predictive accuracy of the model as well as the understanding of variables determining the trajectory, without any knowledge or explanation about the concrete relation between the characteristics of COVID-19 and infection or death rates, as Findl & Suárez argue.[15]

Findl & Suárez identify two problems with de Regt's view on the interrelation between explanation, prediction, and understanding. First, de Regt does not provide any details about how these notions are related, how predictions allow for the refinement of explanations, how explanations enable predictions, and how and where understanding comes in. I agree with Findl & Suárez in this regard. Second, they are not convinced by de Regt's explication of the relation between understanding, explanation, and prediction, which might be due to the first problem identified, the lack of details in de Regt's account. By offering $CIT_1$, de Regt argues that having an intelligible theory, and therefore understanding of that theory, is necessary for generating predictions or characteristic consequences of that theory. Findl & Suárez agree, but they are not convinced that explanation is a necessary intermediate step between intelligibility and prediction, as de Regt argues.[16] Figure 5 depicts the disagreement between de Regt and Findl & Suárez.

*Figure 5: The relations of understanding, explanation and prediction.*[17]

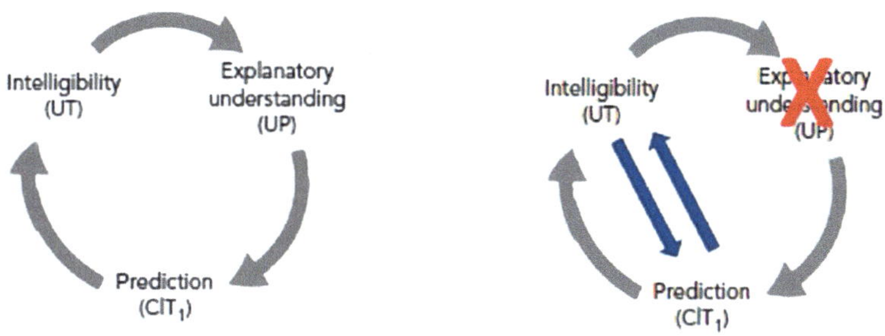


By analyzing the development and use of the model from the Institute of Health Metrics and Evaluation (IHME model), one of the most prominent statistical COVID-19 models at the beginning of the global pandemic in the spring of 2020, Findl & Suárez argue that "[first,] the IHME model satisfies de Regt's intelligibility

15 See ibid. pp. 3f.

16 See ibid. pp. 7f.

17 The chart on the left shows de Regt's view on the relations between understanding, explanation, and prediction (de Regt (2017), p. 108, Fig. 4.1), while the chart on the right displays the criticism put forward by Findl & Suárez (Findl & Suárez (2021), p. 9, Fig. 2).

requirement (i.e., it provides understanding according to [their] terminology) and does so via its predictions; second, that no explanation mediates between intelligibility and predictions (as so-called explanatory understanding would have it), but rather descriptions do."[18]

I am not convinced that Findl & Suárez succeed in arguing that scientific understanding of the dynamics of the COVID-19 pandemic was achieved without explanation. They claim that statistical models provide regularity patterns for a phenomenon, but no causal or counterfactual dependencies, and should, therefore, not be viewed as explanatory.[19] However, bearing in mind my plea to accept an explanatory pluralism in science and my related criticism of narrow accounts of scientific explanation in chapter three, it is important to note that Findl & Suárez also employ an overly restricted notion of explanation. The generic conception of explanation I introduced in section 3.1 requires explanations to provide reasons for the phenomenon to be explained, not necessarily causes. Hence, it can be argued that the assumptions and technical frameworks that are employed by the statistical model and together constitute regularity patterns, or the fit to empirical data provide reasons or are the reasons as to why scientists think that the phenomenon will unfold in a certain manner.

But independently of disagreements concerning the nature of scientific explanation, the work of Findl & Suárez definitely provides important insights for clarifying the relations between understanding and prediction. Especially their finding that predictions were used "backwards", as tests for the assumptions underlying the model and for revising the descriptive understanding already gained at a specific point in time, is crucial for making sense of the role of prediction for understanding. When the model was updated because of incorrect predictions or newly available data, the understanding of the phenomenon gradually improved, too.[20] Hence, Findl & Suárez made a significant contribution to clarifying the relations between understanding and prediction, which can be directly related to issues concerning understanding and representations I pointed out in the previous section. So, there still is much to be learned about how understanding, prediction, explanation, and models are related.

### 7.2.3 The unexplored terrain and the merit of this book

All of these questions concerning the relation of scientific understanding of phenomena to theories, models, representations and prediction, are of course just suggestions in which directions research on scientific understanding might proceed

18 Ibid. p. 16.
19 See ibid. section 4.
20 See ibid. section 5.

from the results of my research project that I consider to be interesting. There are of course plenty of other questions as well. What exactly is grasping? Is understanding always and only an ability possessed by an individual, or can groups of agents have some other kind of understanding as well? Is it satisfying to have an account of understanding that is completely detached from truth? I could extend this list even more, but will leave it like that for the time being. There is still much work to do and many open questions to answer concerning understanding.

I hope that this book provides some helpful guidance and interesting perspectives on how scientists (and subjects generally) come to understand the world. In addressing and answering some of the so far central questions in the philosophical discussion on understanding, this book is a significant contribution in the attempt to resolve existing controversies in the field. By arguing that understanding is an ability that requires knowledge as well as further resources to be manifested, and that understanding manifests in grasping relations and articulating explanations, the GE-account developed in this book consolidates many of the most contested issues related to understanding, and presents a coherent answer how these different concepts are related. In doing so, this book not only offers a new viewpoint on the nature of (scientific) understanding and its relation to knowledge, explanation and scientific practice, but also a starting point to engage with further research questions on understanding in science and also in other contexts.

# Bibliography


Ankeny, R. A. (2000), "Fashioning descriptive models in biology: of worms and wiring diagrams." *Philosophy of Science*, 67, pp. 260–272, DOI: 10.1086/392824.

Ankeny, R. A. & Leonelli, S. (2011), "What's so special about model organisms?" *Studies in History and Philosophy of Science*, 42 (2), pp. 313–323, DOI: 10.1016/j.shpsa.2010.11.039.

Annas, J. (1981), *An Introduction to Plato's Republic*. Oxford, Clarendon Press.

Apel, K.-O. (1982), "The Erklären-Verstehen controversy in the philosophy of the natural and human sciences." In Fløistad G. (ed.), *La philosophie contemporaine / Contemporary philosophy*, International Institute of Philosophy / Institut International de Philosophie, vol 2, pp. 19–49, Dordrecht, Springer, DOI: 10.1007/978-94-010-9940-0_2.

Bailer-Jones, D. (1997), *Scientific Models. A Cognitive Approach with an Application in Astrophysics*, Ph.D. Thesis, University of Cambridge.

Barrow, J. D. (2001), "Cosmology, Life, and the Anthropic Principle". *Annals of the New York Academy of Sciences*, 950 (1), pp. 139–153, DOI: 10.1111/j.1749-6632.2001.tb02133.x.

Baumberger, C. (2011), "Types of Understanding: Their Nature and Their Relation to Knowledge." *Conceptus*, 40, pp. 67–88, DOI: 10.1515/cpt-2014-0002.

Baumberger, C., Beisbart, C. & Brun, G. (2017), "What is Understanding? An Overview of Recent Debates in Epistemology and Philosophy of Science." In Grimm, S., Baumberger, C. & Ammon, S. (eds.), *Explaining Understanding. New Perspectives from Epistemology and Philosophy of Science*, pp. 1–34, New York and London, Routledge.

Baumberger, C. & Brun, G. (2017), "Dimensions of Objectual Understanding." In Grimm S., Baumberger, C. & Ammon, S. (eds.), *Explaining Understanding. New Perspectives from Epistemology and Philosophy of Science*, pp. 165–189, New York and London, Routledge.

Bechtel, W. & Abrahamsen, A. (2005), "Explanation: a mechanist alternative." *Studies in History and Philosophy of Biological and Biomedical Sciences*, 36, pp. 421–441, DOI: 10.1016/j.shpsc.2005.03.010.

Beiser, F. C. (2011), *The German Historicist Tradition*. New York, Oxford University Press, DOI: 10.1093/acprof:oso/9780199691555.001.0001.

Benzer, S. (1968), "Genes and behavior." *Engineering and Science*, 32, pp. 50–52.

Bisgrove, B. W., Essner, J. J. & Yost, H. J. (1999), "Regulation of midline development by antagonism of lefty and nodal signaling." *Development*, 126 (14), pp. 3253–3262, DOI: 10.1242/dev.126.14.3253.

Bohr, N. (1913), "On the Constitution of Atoms and Molecules, Part I". *Philosophical Magazine*, 26 (151), pp. 1–24.

Boyd, K. (2019), "Group understanding." *Synthese*, 198 (7), pp. 6837–6858, DOI: 10.1007/s11229-019-02492-3.

Braillard, P.-A. & Malaterre, C. (2015), *Explanation in Biology. An Enquiry into the Diversity of Explanatory Patterns in the Life Sciences*. In History, Philosophy and Theory of the Life Sciences, Dordrecht, Springer, DOI: 10.1007/978-94-017-9822-8.

Brainard, L. (2020), "How to Explain How-Possibly." *Philosophers Imprint*, 20 (13), pp. 1–23.

Brenner, S. (1974), "The genetics of Caenorhabditis elegans." *Genetics*, 77 (1), pp. 71–94, DOI: 10.1093/genetics/77.1.71.

Brenner, S. (1998), "Letter to Perutz." In Wood, W. B. (ed.), *The nematode Caenorhabditis elegans*, pp. x-xi, Cold Spring Harbor (NY), Cold Spring Harbor Laboratory Press.

Brown, R. J., (1986), "Thought Experiments since the scientific revolution." *International Studies in the Philosophy of Science*, 1(1), pp. 1–15, DOI: 10.1080/02698598608573279.

Brown, J. R. (1991), *The Laboratory of the Mind – Thought Experiments in the Natural Sciences*. New York and London, Routledge.

Cairns, J., Stent, G. S. & Watson, J. D. (eds.) (1966), *Phage and the origins of molecular biology*. Cold Spring Harbor (NY), Cold Spring Harbor Laboratory Press.

Carter, J. & Gordon, E. (2014), "Objectual Understanding and the Value Problem." *American Philosophical Quarterly*, 51 (1), pp. 1–13.

Chang, H. (2004), *Inventing Temperature. Measurement and Scientific Progress*, New York, Oxford University Press, DOI: 10.1093/0195171276.001.0001.

Chang, H. (2012), *Is Water $H_2O$? Evidence, Realism and Pluralism*, Dordrecht, Springer, DOI: 10.1007/978-94-007-3932-1.

Chang, H. (2012), "Beyond Case-Studies: History as Philosophy." In Schmaltz, T. & Mauskopf, S. (eds.) *Integrating History and Philosophy of Science: Problems and Prospects*, (pp. 109–124). Dordrecht, Springer, DOI: 10.1007/978-94-007-1745-9_8.

Chirimuuta, M. (2021), "Prediction versus understanding in computationally enhanced neuroscience." *Synthese*, 199 (1), pp. 767–790, DOI: 10.1007/s11229-020-02713-0.

Choi, S. & Fara, M., "Dispositions", *The Stanford Encyclopedia of Philosophy* (Spring 2021 Edition), Edward N. Zalta (ed.), URL = https://plato.stanford.edu/archives/spr2021/entries/dispositions/ (last accessed April 12$^{th}$, 2022).

Collins, H. (2010), *Tacit and Explicit Knowledge.* Chicago and London, The University of Chicago Press.

Conee, E. & Feldman, R. (2004), *Evidentialism: Essays in Epistemology*. Oxford, Oxford University Press, DOI: 10.1093/0199253722.001.0001.

Contessa, G. (2007), "Scientific Representation, Interpretation, and Surrogative Reasoning", *Philosophy of Science*, 74 (1), pp. 48–68, DOI: 10.1086/519478.

Creaser, C. W. (1934), "The technic of handling the zebra fish (Brachydanio rerio) for the production of eggs which are favorable for embryological research and are available at any specified time throughout the year." *Copeia*, 4, pp. 159–161, DOI: 10.2307/1435845.

Darden, L., (2008), "Thinking Again about Biological Mechanisms." *Philosophy of Science*, 75 (5), pp. 958–969, DOI: 10.1086/594538.

De Campos-Baptista, M. I., Holtzman, N. G., Yelon, D. & Schier, A. F. (2008), "Nodal signaling promotes the speed and directional movement of cardiomyocytes in zebrafish." *Developmental Dynamics: An Official Publication of the American Association of Anatomists*, 237 (12), pp. 3624–3633, DOI: 10.1002/dvdy.21777.

De Regt, H. W. (2004), "Discussion Note: Making Sense of Understanding." *Philosophy of Science*, 71 (1), pp. 98–109, DOI: 10.1086/381415.

De Regt, H. W. (2017), *Understanding Scientific Understanding.* New York, Oxford University Press, DOI: 10.1093/oso/9780190652913.001.0001.

De Regt, H. W. & Gijsbers, V. (2017), "How False Theories Can Yield Genuine Understanding." In Grimm, S., Baumberger, C. & Ammon, S. (eds.), *Explaining Understanding. New Perspectives from Epistemology and Philosophy of Science*, pp. 50–75, New York and London, Routledge.

De Regt, H. W. and Höhl, Anna E. (2020), Review of Khalifa, K., *Understanding, Explanation and Scientific Knowledge*, BJPS Review of Books, https://www.thebsps.org/reviewofbooks/kareem-khalifa-understanding-explanation-and-scientific-knowledge-reviewed-by-de-regt-hohl/ (last accessed April 14th, 2022).

De Vreese, L., Weber, E. & Van Bouwel, J. (2010), "Explanatory pluralism in the medical sciences: Theory and practice." *Theor Med Bioeth*, 31, pp. 371–390, DOI: 10.1007/s11017-010-9156-7.

Dellsén, F. (2020), "Beyond Explanation: Understanding as Dependency Modelling." *British Journal for the Philosophy of Science*, 71, pp. 1261–1286, DOI: 10.1093/bjps/axy058.

Douglas, H. E. (2004), "The Irreducible Complexity of Objectivity." *Synthese*, 138, pp. 453–473, DOI: 10.1023/B:SYNT.0000016451.18182.91.

Driever, W. & Nüsslein-Volhard, C. (1989), "The bicoid protein is a positive regulator of hunchback transcription in the early Drosophila embryo." *Nature*, 337 (6203), pp. 138–143, DOI: 10.1038/337138a0.

Elgin, C. Z. (1991), "Understanding: Art and Science." *Midwest Studies in Philosophy*, 16, pp. 196–208, DOI: 10.1111/j.1475-4975.1991.tb00239.x.

Elgin, C. Z. (2007), "Understanding and the Facts." *Philosophical Studies*, 132, pp. 33–42, DOI: 10.1007/s11098-006-9054-z.

Elgin, C. Z. (2017), *True Enough*. Cambridge (MA) and London, MIT Press.

Endersby, J. (2007), *A guinea pig's history of biology*. London, William Heinemann Ltd.

Fantini, B. (2000), "Molecularizing embryology: Alberto Monroy and the origins of developmental biology in Italy." *The International Journal of Developmental Biology*, 44 (6), pp. 537–553.

Feynman, R. P. (2006 [1985]), *QED: The Strange Theory of Light and Matter*. Princeton, Princeton University Press.

Feynman, R. P. (2017 [1965]), *The Character of Physical Law*. Cambridge (MA), MIT Press.

Findl, J. & Suárez, J. (2021), "Descriptive understanding and prediction in COVID-19 modelling." *History and Philosophy of the Life Sciences*, 43 (4), pp. 1–31, DOI: 10.1007/s40656-021-00461-z.

Fridland, E. & Pavese, C. (eds.) (2021), *The Routledge Handbook of Philosophy of Skills and Expertise*. Routledge.

Friedman, M. (1974), "Explanation and Scientific Understanding." *Journal of Philosophy*, 71 (1), pp. 5–19, DOI: 10.2307/2024924.

Frigg, R. & Nguyen, J. (2016), "The Fiction View of Models Reloaded." *The Monist*, 99, pp. 225–242, DOI: 10.1093/monist/onw002.

Frigg, R. & Nguyen, J., "Scientific Representation", *The Stanford Encyclopedia of Philosophy* (Spring 2020 Edition), Edward N. Zalta (ed.), URL = https://plato.stanford.edu/archives/spr2020/entries/scientific-representation/ (last accessed April 12th, 2022).

Galileo Galilei (1954 [1914, 1638]), *Dialogues concerning two new sciences*. Trans. Crew, H. & de Salvio, A., New York, Dover Publications.

Giere, R. N. (1999), *Science without Laws*. Chicago, University of Chicago Press.

Giere, R. N. (2004), "How models are used to represent reality." *Philosophy of Science*, 71, pp. 742–752, DOI: 10.1086/425063.

Giere, R. N. (2006), *Scientific Perspectivism*. Chicago, University of Chicago Press.

Gijsbers, V. (2013), "Understanding, explanation, and unification." *Studies in the History and Philosophy of Science*, 44, pp. 516–522, DOI: 10.1016/j.shpsa.2012.12.003.

Glymour, B. (2007), "In defence of explanatory deductivism." In Campbell, J. K., O'Rourke, M. & Silverstein, H. (eds.), *Causation and explanation*, pp. 133–154, Cambridge (MA), MIT Press.

Greco, J. (2007), "The Nature of Ability and the Purpose of Knowledge." *Philosophical Issues*, 17, pp. 57–69, DOI: 10.1111/j.1533-6077.2007.00122.x.

Greco, J. (2010), *Achieving Knowledge. A virtue-theoretic account of epistemic normativity*, Cambridge, Cambridge University Press, DOI: 10.1017/CBO9780511844645.

Greco, J. (2014), "Episteme: Knowledge and Understanding." In Timpe, K. & Boyd, C. A. (eds.), *Virtues and Their Vices*, pp. 285–302, Oxford, Oxford University Press, DOI: 10.1093/acprof:oso/9780199645541.003.0014.

Greene, B. (2008), "Put a Little Science in Your Life." New York Times, Open Ed., June 1, https://www.nytimes.com/2008/06/01/opinion/01greene.html (last accessed October 3rd, 2023).

Grimm, S. (2014), "Understanding as Knowledge of Causes." In Fairweather, A. (ed.), *Virtue Epistemology Naturalized: Bridges between Virtue Epistemology and Philosophy of Science*, pp. 347–360, Dordrecht, Springer, DOI: 10.1007/978-3-319-04672-3_19.

Grimm, S. (2017), "Understanding and Transparency." In Grimm, S., Baumberger C. & Ammon, S. (eds.), *Explaining Understanding. New Perspectives from Epistemology and Philosophy of Science*, pp. 212–229, New York and London, Routledge.

Grimm, S., "Understanding", *The Stanford Encyclopedia of Philosophy* (Summer 2021 Edition), Edward N. Zalta (ed.), forthcoming URL = https://plato.stanford.edu/archives/sum2021/entries/understanding/ (last accessed April 11th, 2022).

Gritsman, K., Zhang, J., Cheng, S., Heckscher, E., Talbot, W. S. & Schier, A. F. (1999), "The EGF CFC protein one-eyed pinhead is essential for nodal signaling." *Cell*, 97 (1), pp. 121–132, DOI: 10.1016/s0092-8674(00)80720-5.

Grunwald, D. J., Kimmel, C. B., Westerfield, M., Walker, C. & Streisinger, G. (1988), "A neural degeneration mutation that spares primary neurons in the zebrafish." *Developmental Biology*, 126 (1), pp. 115–128, DOI: 10.1016/0012-1606(88)90245-x.

Hempel, C. G. (1965), *Aspects of Scientific Explanation and Other Essays in the Philosophy of Science*. New York, Free Press.

Hitchcock, C. R. (1999), "Contrastive explanation and the demons of determinism." *British Journal for the Philosophy of Science*, 50 (4), pp. 585–612, DOI: 10.1093/bjps/5 0.4.585.

Hughes, R. I.G. (1997), "Models and Representation", *Philosophy of Science*, 64, pp. 325–336, DOI: 10.1086/392611.

Humphreys, P. (2000), "Analytic Versus Synthetic Understanding." In Fetzer, J. (ed.), *Science, Explanation, and Rationality: The Philosophy of Carl G. Hempel*, pp. 267–286, Oxford, Oxford University Press, DOI: 10.1093/oso/9780199334872.003.0017.

Ingham, P. W. (1997), "Zebrafish genetics and its implications for understanding vertebrate development." *Human Molecular Genetics*, 6 (10), pp. 1755–1760, DOI: 10.1093/hmg/6.10.1755.

Khalifa, K. (2013), "Is understanding explanatory or objectual?" *Synthese*, 190, pp. 1153–1171, DOI: 10.1007/s11229-011-9886-8.

Khalifa, K. (2017a), "Must Understanding be Coherent?" In Grimm, S., Baumberger, C. & Ammon, S. (eds.), *Explaining Understanding. New Perspectives from Epistemology and Philosophy of Science*, pp. 139–164, New York and London, Routledge.

Khalifa, K. (2017b), *Understanding, Explanation, and Scientific Knowledge*. Cambridge, Cambridge University Press, DOI: 10.1017/9781108164276.

Kim, J. (1974), "Noncausal Connections." *Noûs*, 8, pp. 41–52, DOI: 10.2307/2214644.

Kim, J. (1994 [2010]), "Explanatory Knowledge and Metaphysical Dependence." *Philosophical Issues*, 5, pp. 51–69, Reprinted in Kim, J. (2010), *Essays in the Metaphysics of Mind*. New York, Oxford University Press, DOI: 10.1093/acprof:oso/9780199585878.001.0001.

Kimmel, C. B., Warga, R. M. & Schilling, T. F. (1990) "Origin and organization of the zebrafish fate map." *Development*, 108 (4), pp. 581–594, DOI: 10.1242/dev.108.4.581.

Kimmel, C. B., Ballard, W. W., Kimmel, S. R., Ullmann, B. & Schilling, T. F. (1995), "Stages of embryonic development of the zebrafish." *Developmental Dynamics: An Official Publication of the American Association of Anatomists*, 203 (3), pp. 253–310, DOI: 10.1002/aja.1002030302.

Kitcher, P. & Salmon, W. C. (1987), "Van Fraassen on explanation." *Journal of Philosophy*, 84 (6), pp. 315–330, DOI: 10.2307/2026782.

Kitcher, P. (1989), "Explanatory Unification and the Causal Structure of the World." In Kitcher, P. & Salmon, W. (eds.), *Scientific Explanation*, Minnesota Studies in the Philosophy of Science, Vol. 13, pp. 410–505, Minneapolis (MN), University of Minnesota Press.

Knorr-Cetina, K. (1999), *Epistemic Cultures: How the Sciences Make Knowledge*. Cambridge (MA), Harvard University Press, DOI: 10.4159/9780674039681.

Kragh, H. (2003), "Magic Number: A Partial History of the Fine-Structure Constant." *Archive for History of Exact Sciences*, 57 (5), pp. 395–431, DOI: 10.1007/s00407-002-0065-7.

Kuhn, T. S. (2012 [1970]), *The structure of scientific revolutions* (4. ed., 50th anniversary ed.). Chicago, University of Chicago Press.

Kühne, U. (1997), "Gedankenexperiment und Erklärung." *Bremer Philosophica*, 5, pp. 1–51.

Kvanvig, J. L. (2003), *The Value of Knowledge and the Pursuit of Understanding*. New York, Cambridge University Press, DOI: 10.1017/CBO9780511498909.

Kvanvig, J. L. (2009), "The value of understanding." In Haddock, A., Miller, A. & Pritchard, D. (eds.), *Epistemic value*, pp. 95–111, Oxford, Oxford University Press, DOI: 10.1093/acprof:oso/9780199231188.001.0001.

Laale, H. W. (1977), "The biology and use of zebrafish, Brachydanio rerio in fisheries research. A literature review." *Journal of Fish Biology*, 10, pp. 121–173, DOI: 10.1111/j.1095-8649.1977.tb04049.x.

Lear, J. (1988), *Aristotle: The Desire to Understand*. New York, Cambridge University Press, DOI: 10.1017/CBO9780511570612.002.

Leonelli, S. (2009), "Understanding in Biology: The Impure Nature of Biological Knowledge." In de Regt, H. W., Leonelli, S. & Eigner, K. (eds.), *Understanding: Philosophical Perspectives*, pp. 189–209, Pittsburgh, University of Pittsburgh Press.

Lipton, P. (2009), "Understanding without explanation." In de Regt, H. W., Leonelli, S. & Eigner, K. (eds.), *Scientific Understanding: Philosophical Perspectives*, pp. 43–63, Pittsburgh, University of Pittsburgh Press.

Longino, H. E. (1990), *Science as Social Knowledge: Values and Objectivity in Scientific Inquiry*. Princeton (NJ), Princeton University Press.

MacBride, F., "Relations", *The Stanford Encyclopedia of Philosophy* (Winter 2020 Edition), Edward N. Zalta (ed.), URL = https://plato.stanford.edu/archives/win2020/entries/relations/ (last accessed April 14th, 2022).

Machamer, P., Darden, L. & Craver, C. F. (2000), "Thinking about Mechanisms." *Philosophy of Science*, 67 (1), pp. 1–25, DOI: 10.1086/392759.

Maier, J., "Abilities", *The Stanford Encyclopedia of Philosophy* (Summer 2021 Edition), Edward N. Zalta (ed.), URL = https://plato.stanford.edu/archives/sum2021/entries/abilities/ (last accessed April 12th, 2022).

Makkreel, R., "Wilhelm Dilthey", *The Stanford Encyclopedia of Philosophy* (Spring 2021 Edition), Edward N. Zalta (ed.), URL = https://plato.stanford.edu/archives/spr2021/entries/dilthey/ (last accessed April 11th, 2022).

Mantzavinos, C. (2016), *Explanatory Pluralism*. Cambridge, Cambridge University Press, DOI: 10.1017/CBO9781316440599.

Meinefeld, W. (1995), *Realität und Konstruktion. Erkenntnistheoretische Grundlagen einer Methodologie der empirischen Sozialforschung*, Wiesbaden, VS Verlag für Sozialwissenschaften, DOI: 10.1007/978-3-663-11243-3.

Meunier, R. (2012), "Stages in the development of a model organism as a platform for mechanistic models in developmental biology: Zebrafish, 1970–2000." *Studies in History and Philosophy of Biological and Biomedical Sciences*, 43, pp. 522–531, DOI: 10.1016/j.shpsc.2011.11.013.

Mullins, M. C., Hammerschmidt, M., Haffter, P. & Nüsslein-Volhard, C. (1994), "Large-scale mutagenesis in the zebrafish: in search of genes controlling development in a vertebrate." *Current Biology*, 4 (3), pp. 189–202, DOI: 10.1016/s0960-9822(00)00048-8.

Newman, M. (2017), "An Evidentialist Account of Explanatory Understanding." In Grimm, S., Baumberger, C. & Ammon, S. (eds.), *Explaining Understanding. New Perspectives from Epistemology and Philosophy of Science*, pp. 190–211, New York and London, Routledge.

Nüsslein-Volhard, C. & Wieschaus, E. (1980), "Mutations affecting segment number and polarity in Drosophila." *Nature*, 287 (5785), pp. 795–801, DOI: 10.1038/287795a0.

Oberheim, E. & Hoyningen-Huene, P., "The Incommensurability of Scientific Theories", *The Stanford Encyclopedia of Philosophy* (Fall 2018 Edition), Edward N. Zalta (ed.), URL = https://plato.stanford.edu/archives/fall2018/entries/incommensurability/ (last accessed April 16th, 2022)

Olsson, E., "Coherentist Theories of Epistemic Justification", *The Stanford Encyclopedia of Philosophy* (Fall 2021 Edition), Edward N. Zalta (ed.), URL = https://plato.stanford.edu/archives/fall2021/entries/justep-coherence/ (last accessed April 11th, 2022).

Pappas, G., "Internalist vs. Externalist Conceptions of Epistemic Justification", *The Stanford Encyclopedia of Philosophy* (Fall 2017 Edition), Edward N. Zalta (ed.), URL = https://plato.stanford.edu/archives/fall2017/entries/justep-intext/ (last accessed April 11th, 2022).

Pasnau, R. (2017), *After Certainty: A History of Our Epistemic Ideals and Illusions*. New York, Oxford University Press, DOI: 10.1093/oso/9780198801788.001.0001.

Pavese, C., "Knowledge How", *The Stanford Encyclopedia of Philosophy* (Summer 2021 Edition), Edward N. Zalta (ed.), URL = https://plato.stanford.edu/archives/sum2021/entries/knowledge-how/ (last accessed April 12th, 2022).

Polanyi, M. (1962 [1958]), *Personal Knowledge. Towards a Post-Critical Philosophy*, London, Routledge.

Potochnik, A. (2017), *Idealization and the Aims of Science*, Chicago (IL), University of Chicago Press, DOI: 10.7208/9780226507194.

Pritchard, D., Millar, A. & Haddock, A. (2010), *The Nature and Value of Knowledge: Three Investigations*, Oxford, Oxford University Press, DOI: 10.1093/acprof:oso/9780199586264.001.0001.

Pritchard, D. (2010), "Knowledge and Understanding." In Pritchard, D., Millar, A. & Haddock, A., *The Nature and Value of Knowledge: Three Investigations*, (pp. 3–90), New York, Oxford University Press, DOI: 10.1093/acprof:oso/9780199586264.001.0001.

Railton, P. (1978), "A Deductive-Nomological Model of Probabilistic Explanation." *Philosophy of Science*, 45 (2), pp. 206–226, DOI: 10.1086/288797.

Reese, H. W. (1999), "Explanation Is Not Description." *Behavioral Development Bulletin*, 8 (1), pp. 3–7, DOI: 10.1037/h0100524.

Reutlinger, A., Hangleiter, D. & Hartmann, S. (2018). „Understanding (with) Toy Models." *British Journal for the Philosophy of Science*, 69 (4), pp. 1069–1099, DOI: 10.1093/bjps/axx005.

Reutlinger, A., & Saatsi, J. (2018). *Explanation beyond causation: philosophical perspectives on noncausal explanations*. Oxford, Oxford University Press, DOI: 10.1093/oso/9780198777946.001.0001.

Reydon, T. (2012), "How-possibly explanations as genuine explanations and helpful heuristics: A comment on Forber." *Studies in History and Philosophy of Biological and Biomedical Sciences*, 43 (1), pp. 302–310, DOI: 10.1016/j.shpsc.2011.10.015.

Rheinberger, H.-J. (1997), *Toward a history of epistemic things. Synthesizing proteins in the test tube*, Stanford (CA), Stanford University Press.

Ribeiro, R. & Collins, H. (2007), "The Bread- Making Machine, Tacit Knowledge, and Two Types of Action." *Organization Studies,* 28 (9), pp. 1417–1433, DOI: 10.1177/0170840607082228.

Rice, C. & Rohwer, Y. (2021), "How to Reconcile a Unified Account of Explanation with Explanatory Diversity." *Found Sci,* 26, pp. 1025–1047, DOI: 10.1007/s10699-019-09647-y.

Robertson, D. S. (1996), "Niels Bohr – Through Hydrogen Towards the Nature of Matter." In Lakhtakia, A. (ed.), *Models and Modelers of Hydrogen*, pp. 49–82, Singapore, World Scientific Publishing.

Rudin, C., et al. (2021). "Interpretable machine learning: Fundamental principles and 10 grand challenges." *Statistics Surveys*, 16, pp. 1–85, DOI: 10.1214/21-SS133.

Ryle, G. (1949), *The Concept of Mind*. Chicago, The University of Chicago Press.

Ryle, G. (1990 [1946]), "Knowing How and Knowing That." In Collected Papers (Volume 2), Bristol, Thoemmes Antiquarian Books Ltd, pp. 212–225.

Sakaguchi, K. et. al. (2019), "Comprehensive Experimental System for a Promising Model Organism Candidate for Marine Teleosts." *Scientific Reports*, 9 (4948), DOI : 10.1038/s41598-019-41468-8.

Schier, A. F., Neuhauss, S. C., Helde, K. A., Talbot, W. S. & Driever, W. (1997), "The one-eyed pinhead gene functions in mesoderm and endoderm formation in zebrafish and interacts with no tail." *Development,* 124 (2), pp. 327–342, DOI: 10.1242/dev.124.2.327.

Schmaltz, T. & Mauskopf, S. (eds.) (2012), *Integrating History and Philosophy of Science: Problems and Prospects*. Dordrecht, Springer, DOI: 10.1007/978-94-007-1745-9.

Soler, L., Zwart, S., Lynch, M. & Israël-Jost, V. (eds.) (2014), *Science After the Practice Turn in the Philosophy, History, and Social Studies of Science*. Studies in the Philosophy of Science 14, New York and London, Routledge.

Sosa, E. (2010), "How competence matters in epistemology." *Philosophical Perspectives*, 24, pp. 465–475, DOI: 10.1111/j.1520-8583.2010.00200.x.

Stahl, F. W. (1995), "George Streisinger—December 27, 1927—September 5, 1984." *Biographical Memoirs. National Academy of Sciences (U.S.)*, 68, pp. 353–361.

Streisinger, G., Walker, C., Dower, N., Knauber, D. & Singer, F. (1981), "Production of clones of homozygous diploid zebra fish (Brachydanio rerio)." *Nature*, 291 (5813), pp. 293–296, DOI: 10.1038/291293a0.

Strevens, M. (2008), *Depth*. Cambridge (MA) and London, Harvard University Press.

Strevens, M. (2013), "No Understanding without Explanation", *Studies in History and Philosophy of Science A*, 44 (3), pp. 510–515, DOI: 10.1016/j.shpsa.2012.12.005.

Strevens, M. (2017), "How Idealizations Provide Understanding." In Grimm, S., Baumberger, C. & Ammon, S. (eds.), *Explaining Understanding. New Perspectives from Epistemology and Philosophy of Science*, pp. 37–49, New York and London, Routledge.

Sullivan, E. (2022), "Understanding from machine learning models." *The British Journal for the Philosophy of Science*, 73(1), pp. 109–133, DOI: 10.1093/bjps/axz035.

Trout, J. D. (2002), "Scientific Explanation And The Sense Of Understanding." *Philosophy of Science*, 69 (2), pp. 212–233, DOI: 10.1086/341050.

Turri, J., Alfano, M. & Greco, J., "Virtue Epistemology", *The Stanford Encyclopedia of Philosophy* (Winter 2021 Edition), Edward N. Zalta (ed.), URL = https://plato.stanford.edu/archives/win2021/entries/epistemology-virtue/ (last accessed April 12th 2022).

Uebel, T. (2010), "Opposition to *Verstehen* in Orthodox Logical Empiricism." In Feest, U. (ed.), *Historical Perspectives on Erklären and Verstehen*, pp. 291–309, Dordrecht, Springer, DOI: 10.1007/978-90-481-3540-0_15.

Van Fraassen, B. C. (1980), *The Scientific Image*. Oxford, Clarendon Press, DOI:10.1093/0198244274.001.0001.

Várlaki, P., Nádai, L., Bokor, J. (2008), "Number archetypes and 'background' control theory concerning the fine structure constant". *Acta Polytechnica Hungarica*, 5 (2), pp. 71–104.

Varzi, A., "Mereology", *The Stanford Encyclopedia of Philosophy* (Spring 2019 Edition), Edward N. Zalta (ed.), URL = https://plato.stanford.edu/archives/spr2019/entries/mereology/ (last accessed April 14th, 2022).

Vetter, B. (2019), "Are abilities dispositions?" *Synthese*, 196, pp. 201–220, DOI: 10.1007/s11229-016-1152-7.

Weber, E., de Regt, H. W. & van Eck, D. (2021), "Investigating the Unity and Disunity of Scientific Explanation." Found Sci, 26, pp. 1021–1024, DOI: 10.1007/s10699-020-09704-x.

Weisberg, M. (2013), *Simulation and Similarity. Using models to understand the world*, Oxford, Oxford University Press, DOI: 10.1093/acprof:oso/9780199933662.001.0001.

Whittaker, E. (1945), "Eddington's Theory of the Constants of Nature." *The Mathematical Gazette*, 29 (286), pp. 137–144, DOI: 10.2307/3609461.

Wilkenfeld, D. A. (2013), "Understanding as Representation Manipulability." *Synthese*, 190 (6), pp. 997–1016, DOI: 10.1007/s11229-011-0055-x.

Winther, R. G., "The Structure of Scientific Theories", *The Stanford Encyclopedia of Philosophy* (Spring 2021 Edition), Edward N. Zalta (ed.), URL = https://plato.stanford.edu/archives/spr2021/entries/structure-scientific-theories/ (last accessed April 14th, 2022).

Woodward, J. (2003), *Making things happen: a theory of causal explanation*. New York, Oxford University Press, DOI: 10.1093/0195155270.001.0001.

Woodward, J. & Ross, L., "Scientific Explanation", *The Stanford Encyclopedia of Philosophy* (Summer 2021 Edition), Edward N. Zalta (ed.), URL = https://plato.stanford.edu/archives/sum2021/entries/scientific-explanation/ (last accessed April 12th, 2022).

Wright, C. & van Eck, D. (2018), "Ontic Explanation Is either Ontic or Explanatory, but Not Both." *Ergo: An Open Access Journal of Philosophy*, 5, pp. 997–1029, DOI: 10.3998/ergo.12405314.0005.038.

Zagzebski, L. (1996), *Virtues of the Mind. An Inquiry into the Nature of Virtue and the Ethical Foundations of Knowledge*, New York, Cambridge University Press, DOI: 10.1017/CBO9781139174763.

Zagzebski, L. (2001), "Recovering Understanding." In Steup, M. (ed.), *Knowledge, Truth, and Duty: Essays on Epistemic Justification, Responsibility, and Virtue*, pp. 235–252, New York, Oxford University Press, DOI: 10.1093/0195128923.003.0015.

Zhang, J., Talbot, W. S. & Schier, A. F. (1998), "Positional cloning identifies zebrafish one-eyed pinhead as a permissive EGF-related ligand required during gastrulation." *Cell*, 92 (2), pp. 241–251, DOI: 10.1016/S0092-8674(00)80918-6.